Review of the St. Johns River Water Supply Impact Study: Report 1

Committee to Review the St. Johns River Water Supply Impact Study

Water Science and Technology Board

Division on Earth and Life Studies

NATIONAL RESEARCH COUNCIL
OF THE NATIONAL ACADEMIES

THE NATIONAL ACADEMIES PRESS
Washington, D.C.
www.nap.edu

THE NATIONAL ACADEMIES PRESS   500 Fifth Street, N.W.   Washington, DC 20001

NOTICE: The project that is the subject of this report was approved by the Governing Board of the National Research Council, whose members are drawn from the councils of the National Academy of Sciences, the National Academy of Engineering, and the Institute of Medicine. The members of the panel responsible for the report were chosen for their special competences and with regard for appropriate balance.

Support for this study was provided by the St. Johns River Water Management District under grant SLOC-25123. Any opinions, findings, conclusions, or recommendations expressed in this publication are those of the author(s) and do not necessarily reflect the views of the organizations or agencies that provided support for the project.

International Standard Book Number 13: 978-0-309-14222-9
Library of Congress Catalog Card Number 10: 0-309-14222-9

Additional copies of this report are available from the National Academies Press, 500 5th Street, N.W., Lockbox 285, Washington, DC 20055; (800) 624-6242 or (202) 334-3313 (in the Washington metropolitan area); Internet, http://www.nap.edu.

Copyright 2009 by the National Academy of Sciences. All rights reserved.

Printed in the United States of America.

THE NATIONAL ACADEMIES
Advisers to the Nation on Science, Engineering, and Medicine

The **National Academy of Sciences** is a private, nonprofit, self-perpetuating society of distinguished scholars engaged in scientific and engineering research, dedicated to the furtherance of science and technology and to their use for the general welfare. Upon the authority of the charter granted to it by the Congress in 1863, the Academy has a mandate that requires it to advise the federal government on scientific and technical matters. Dr. Ralph J. Cicerone is president of the National Academy of Sciences.

The **National Academy of Engineering** was established in 1964, under the charter of the National Academy of Sciences, as a parallel organization of outstanding engineers. It is autonomous in its administration and in the selection of its members, sharing with the National Academy of Sciences the responsibility for advising the federal government. The National Academy of Engineering also sponsors engineering programs aimed at meeting national needs, encourages education and research, and recognizes the superior achievements of engineers. Dr. Charles M. Vest is president of the National Academy of Engineering.

The **Institute of Medicine** was established in 1970 by the National Academy of Sciences to secure the services of eminent members of appropriate professions in the examination of policy matters pertaining to the health of the public. The Institute acts under the responsibility given to the National Academy of Sciences by its congressional charter to be an adviser to the federal government and, upon its own initiative, to identify issues of medical care, research, and education. Dr. Harvey V. Fineberg is president of the Institute of Medicine.

The **National Research Council** was organized by the National Academy of Sciences in 1916 to associate the broad community of science and technology with the Academy's purposes of furthering knowledge and advising the federal government. Functioning in accordance with general policies determined by the Academy, the Council has become the principal operating agency of both the National Academy of Sciences and the National Academy of Engineering in providing services to the government, the public, and the scientific and engineering communities. The Council is administered jointly by both Academies and the Institute of Medicine. Dr. Ralph J. Cicerone and Dr. Charles M. Vest are chair and vice chair, respectively, of the National Research Council.

www.national-academies.org

COMMITTEE TO REVIEW THE ST. JOHNS RIVER WATER SUPPLY IMPACT STUDY

PATRICK L. BREZONIK, *Chair*, University of Minnesota, Minneapolis
M. SIOBHAN FENNESSY, Kenyon College, Gambier, Ohio
BEN R. HODGES, University of Texas, Austin
JAMES R. KARR, University of Washington, Seattle
MARK S. PETERSON, University of Southern Mississippi, Ocean Springs
JAMES L. PINCKNEY, University of South Carolina, Columbia
JORGE I. RESTREPO, Florida Atlantic University, Boca Raton
ROLAND C. STEINER, Washington Suburban Sanitary Commission, Laurel, Maryland
J. COURT STEVENSON, University of Maryland, Cambridge

NRC STAFF

LAURA J. EHLERS, Study Director
STEPHANIE E. JOHNSON, Interim Study Director (February 2009 to June 2009)
MICHAEL J. STOEVER, Project Assistant

WATER SCIENCE AND TECHNOLOGY BOARD

CLAIRE WELTY, *Chair*, University of Maryland, Baltimore County
JOAN G. EHRENFELD, Rutgers University, New Brunswick, New Jersey
GERALD E. GALLOWAY, University of Maryland, College Park
SIMON GONZALEZ, National Autonomous University of Mexico, Mexico City
CHARLES N. HAAS, Drexel University, Philadelphia, Pennsylvania
KENNETH R. HERD, Southwest Florida Water Management District, Brooksville
JAMES M. HUGHES, Emory University, Atlanta, Georgia
THEODORE L. HULLAR, Cornell University, Ithaca, New York
KIMBERLY L. JONES, Howard University, Washington, DC
G. TRACY MEHAN, The Cadmus Group, Inc., Arlington, Virginia
DAVID H. MOREAU, University of North Carolina, Chapel Hill
THOMAS O' ROURKE, Cornell University, Ithaca, New York
DONALD I. SIEGEL, Syracuse University, Syracuse, New York
SOROOSH SOROOSHIAN, University of California, Irvine
HAME M. WATT, Independent Consultant, Washington, DC
JAMES L. WESCOAT, JR., Massachusetts Institute of Technology, Cambridge

STAFF

STEPHEN D. PARKER, Director
JEFFREY W. JACOBS, Scholar
LAURA J. EHLERS, Senior Program Officer
STEPHANIE E. JOHNSON, Senior Program Officer
LAURA E. HELSABECK, Associate Program Officer
M. JEANNE AQUILINO, Financial and Administrative Associate
ELLEN A. DE GUZMAN, Senior Program Associate
ANITA A. HALL, Senior Program Associate
MICHAEL J. STOEVER, Senior Program Assistant
STEPHEN T. RUSSELL, Senior Program Assistant

Acknowledgment of Reviewers

This report has been reviewed in draft form by individuals chosen for their diverse perspectives and technical expertise, in accordance with procedures approved by the National Research Council's Report Review Committee. The purpose of this independent review is to provide candid and critical comments that will assist the institution in making its published report as sound as possible and to ensure that the report meets institutional standards for objectivity, evidence, and responsiveness to the study charge. The review comments and draft manuscript remain confidential to protect the integrity of the deliberative process. We wish to thank the following individuals for their review of this report:

EMILY S. BERNHARDT, Duke University
MARK M. BRINSON, East Carolina University
WENDY D. GRAHAM, University of Florida
MICHAEL C. KAVANAUGH, Malcolm Pirnie, Inc.
JUDITH L. MEYER, University of Georgia
ERNST B. PEEBLES, University of South Florida

Although the reviewers listed above have provided many constructive comments and suggestions, they were not asked to endorse the conclusions or recommendations nor did they see the final draft of the report before its release. The review of this report was overseen by **Jerome B. Gilbert**. Appointed by the National Research Council, he was responsible for making certain that an independent examination of this report was carried out in accordance with institutional procedures and that all review comments were carefully considered. Responsibility for the final content of this report rests entirely with the authoring committee and the institution.

Contents

SUMMARY		1
1	INTRODUCTION	10
	SJRWMD Water Resources Planning, *13*	
	Water Supply Impact Study, *15*	
	NRC Study and Report Roadmap, *16*	
2	ST. JOHNS RIVER WATER SUPPLY IMPACT STUDY: CROSS-CUTTING ISSUES	18
	Study Focus, *19*	
	Need for Integration, *20*	
	Water and Nutrient Budgets, *24*	
	Dewatering of Floodplains and Wetlands, *25*	
	Summary, *26*	
3	HYDRODYNAMIC AND HYDROLOGIC MODELING	27
	Surface Water Hydrodynamics and Hydrology, *27*	
	Groundwater Hydrology, *35*	
4	ECOLOGICAL ASSESSMENTS	43
	Biogeochemistry, *43*	
	Plankton, Nutrients, and TMDLs, *47*	
	Benthos, *51*	
	Littoral Zone, *54*	
	Fish, *57*	
	Wetlands and Wetland-Dependent Species, *64*	
REFERENCES		72
APPENDIXES		
A	Acronyms	80
B	Biographical Sketches for Committee to Review the St. Johns River Water Supply Impact Study	82

Summary

The St. Johns River is the longest river in Florida, flowing 310 miles from Indian River County north to the Atlantic Ocean. The river drops only 30 feet from its headwaters to its mouth, such that the river has extensive freshwater wetlands, numerous large lakes, a wide estuarine channel, and a correspondingly diverse array of native flora and fauna. Water resource management in the river's watershed, which accounts for 23 percent of Florida's land area, is the responsibility of the St. Johns River Water Management District (the District). The District must provide water for the region's 4.4 million residents as well as numerous industrial and agricultural users, all while protecting natural systems within the river basin.

With population growth in the watershed expected to surpass 7.2 million in 2030, the District, through its water resources planning process, has begun to identify alternative sources of water beyond its traditional groundwater sources. Water reuse, desalination, and new surface water supplies are all under consideration, including the potential withdrawal of 262 million gallons per day (MGD) from the St. Johns River. To more comprehensively evaluate the environmental impacts of withdrawing this water from the river, in early 2008 the District embarked on a two-year Water Supply Impact Study (WSIS). Later that year, the District requested the involvement of the National Research Council (NRC) to review scientific aspects of the study and provide advice to its ongoing effort (see Chapter 1 for the NRC committee's statement of task). This first report of the committee reviews the Phase I work of the WSIS and provides recommendations for improving Phase II. The report is organized along the lines of the seven scientific workgroups of the District, which include hydrologic and hydrodynamic modeling, biogeochemistry, plankton and water quality, benthos, the littoral zone, fish, and wetlands and wetland-dependent species.

It should be noted that the WSIS does not consider the largest tributary to the St. Johns, the Ocklawaha River, to the same degree as the St. Johns River itself because (1) the Ocklawaha is hydrologically distinct from the St. Johns, (2) a focused study on the St. Johns River was more feasible given time and resource constraints, and (3) a separate analysis of minimum flow and level requirements for the Ocklawaha River is planned for the near future. Additionally, the WSIS does not consider the impacts of population growth that an additional water withdrawal of 262 MGD could support, as the District has no direct authority over growth and associated land use changes in the basin.

OVERARCHING ISSUES

In its assessment of the WSIS, the committee found that several issues transcended the discipline-based topics of the individual workgroups. Three are discussed here, with additional issues found in Chapter 2. The committee feels that these should be the highest priority issues for the District to consider as Phase II commences. Directing resources towards these issues now (as opposed to some of the individual projects felt to be of less importance—see Chapter 4 for details) would be in the best interests of the District in order to fully understand the effects of the proposed water withdrawals.

Integration

There is a substantial need for integration among workgroups as the WSIS proceeds, such that the output from one workgroup serves as input to another group for an iterative analysis of ecosystem impacts. Although results from the hydrodynamic and hydrologic modeling are being readily incorporated into the plans of the other workgroups, it is not clear that information exchange goes in the opposite direction or that the ecological workgroups are sufficiently linked to one another. For example, changes in the benthic macroinvertebrate assemblage may engender changes in the fish assemblage. Oral discussions with the District indicate recognition of this issue, but the Phase I report does not provide much evidence for integrative and cross-workgroup analyses.

One way of facilitating integration of the seven workgroups would be for the District to develop a conceptual framework of qualitative interactions that link various ecological and physical parameters affected by surface water withdrawals. Not only would such a framework be an important communication tool for the District, but it could also be used to better connect the field studies, models, and analyses that are part of the WSIS. Along with this framework, the District might provide a clearer set of testable hypotheses and quantitative research questions that link the hydrodynamics and hydrology of the WSIS to the ecology and reflect the state of knowledge along with the planned studies. That is, the proposed conceptual framework described above provides organizing principles for qualitative understanding of the linkages, whereas the hypotheses and research questions provide the guidelines for the scientific efforts required to quantify the linkages.

Finally, several workgroups of the WSIS are analyzing the potential impacts of additional water withdrawals on individual indicator species of submersed aquatic vegetation (SAV), benthos, fish, and birds—an approach that could hinder the ability of the District to achieve integration. It generally is not clear whether the most sensitive species are being considered. Moreover, reliance on single species as indicators of change may underestimate community-level impacts and may ultimately limit the District's ability to understand a variety of important ecological dynamics caused by water withdrawals.

Water and Nutrient Budgets and Return Flows

It will be important for the District to assemble basic water and nutrient budget information for the St. Johns River basin and its major subunits. These budgets should include the standard sources, sinks, and storage components of such budgets, as well as information on how these components vary over time. The purpose of such budgets is primarily to better understand the proposed water withdrawals relative to current conditions in the St. Johns River. The District estimates that the proposed continuous withdrawal at DeLand, Florida, would constitute approximately 7.8 percent of the average daily flow over the period of record. This is a significant fraction that would increase under low flow conditions.

Nutrient budgets are needed to conduct a credible analysis of the impacts of water withdrawals on algal blooms in the river and especially in the large lakes that constitute major parts of middle portion of the St. Johns River. Comprehensive information on nitrogen and phosphorus loadings to the system currently is lacking or at least has not been assembled in a system-wide way.

Finally, potential water withdrawals are treated in the WSIS as consumptive uses with no return flows of withdrawn water to the system—a conservative assumption according to the District. Although it is difficult to estimate future patterns of consumptive water use, given uncertainties regarding future climatic and land-use conditions, data are available on past and current consumptive and non-consumptive uses of water in the drainage basin for the District to be able to provide a range of scenarios of the extent to which future withdrawals will be consumptive.

Dewatering of Floodplains and Wetlands

The Phase I hydrodynamic and hydrologic modeling studies suggest that the decline in surface water levels produced by the proposed water withdrawal will be small, from 1 cm to at most 4 cm. This result has played a role in guiding the Phase I work of the biogeochemistry, SAV, benthos, fish, and wetlands workgroups. Although initial consideration of this result might suggest that the likely environmental effects of water withdrawal will be minimal, the District should avoid a rush to judgment. First, more careful analysis in the form of advanced hydrologic and wetland modeling is needed to determine the area to be dewatered as a result of lower water levels, including the timing and duration of dewatering events. This type of modeling will be a part of the Phase II studies of the hydrodynamics and hydrologic and wetlands workgroups. In addition, an effort should be made during Phase II to determine the nature and areal extent of locations that may experience altered biology due to *partial* dewatering of wetlands and floodplains. Indeed, gaining more definitive information on the areal extent of dewatering would be more immediately valuable to understanding the environmental impacts of water withdrawal than several of the literature and monitoring studies suggested for Phase II (see Chapter 4 for details).

HYDRODYNAMIC AND HYDROLOGICAL MODELING

Extensive surface water and groundwater modeling was conducted as part of Phase I of the WSIS. The surface water hydrodynamic and hydrological studies focused on understanding the changes in surface water depth, discharge, water age, salinity, turbidity, and wetland dewatering caused by surface water withdrawals from the St. Johns River. The Phase I studies were screening studies based on readily available historical data and hydrological models. The workgroup developed hydrodynamic models of the lower and middle St. Johns River, it evaluated salinity under different scenarios for the lower river, and it conducted a model analysis of possible changes in sediment loading in the middle river. They concluded that water stage in the lower and middle river will be relatively unchanged by surface water withdrawals. Channel dredging and surface water withdrawals were predicted to increase salinity in the lower St. Johns River more than rising sea level and wastewater diversion. However, the analysis focused on increases in average salinity and did not discuss extremes. It will be important for the Phase II modeling and analyses to be tied carefully to the time–space scales of salinity that are important to ecology; indeed, the hydrodynamic modelers should have a documented process for determining scenarios and data needs for the other six workgroups.

Overall, the District is progressing along the correct track with respect to the surface water studies, but critical details either have not been considered or have not been sufficiently documented. Specifically, the District should work to connect the separate modeling and analysis efforts of Phase I. For example, the Phase I analyses did not examine how hydrologic changes in specific river sections relate to changes or effects in the rest of the river. Second, several areas were not studied or explained in enough detail during Phase I, including modeling of extreme conditions, the hydrology of the upper St. John River, vertical gradients of salinity and horizontal upstream salinity excursions in the lower river, and the effects of bridges and sea level rise. Third, the District should document model calibration and sensitivity and develop methods to quantify model uncertainty on the time–space scales at which ecological effects occur.

The primary goal of the Phase I groundwater modeling was to predict whether discharges of groundwater into the St. Johns River would change if the river stage dropped due to water withdrawals. Two groundwater flow models were used during Phase I to compute groundwater base flows along the river from the surficial aquifer system and the upper Floridan aquifer: the North Central Florida and the East Central Florida MODFLOW models. The groundwater analyses were confined to the middle and upper St. Johns River basins. Based on the modeling results, the change in the average discharge of groundwater following water withdrawal was not predicted to be particularly significant in the middle and upper basins, although the change in chloride flux was predicted to be significant. This conclusion may be correct, but it is not yet technically defensible due to limitations in the models.

In order to improve the groundwater modeling in Phase II of the WSIS, the District should consider using a transient model that includes wetlands processes (e.g., the East-Central Florida Transient [ECFT] model) and a cross-sectional density-dependent model. ECFT would be a valuable screening tool to better understand how groundwater flow into the river will vary with water withdrawals due to its representation of various seasonal changes, wetland simulation capabilities, and increase in horizontal resolution. A two-dimensional, cross-sectional, density-dependent model could be developed to indicate the vulnerability of certain river segments to changes in saline flow from groundwater following potential water withdrawals from the river or other alternative water management scenarios. This would be most valuable in regions of the river known to be susceptible to saltwater intrusion.

ECOLOGICAL WORKGROUPS

In the Phase I study, six workgroups used hydrologic modeling data, existing monitoring data, and literature reviews to provide preliminary assessments of potential biogeochemical and ecological impacts in the St. Johns River from withdrawing 262 MGD of surface water. Chapter 4 provides assessments of the Phase I work on biogeochemistry; plankton, nutrients, and total maximum daily loads (TMDLs); the littoral zone; benthos; fish; and wetlands and wetland-dependent species. This summary includes only a subset of the concerns held by the committee with regard to each workgroup. Additional detail can be found in the individual sections of Chapter 4.

Biogeochemistry

The biogeochemistry workgroup identified seven potential effects of additional water withdrawals on biogeochemical processes, all related to the possibility that soil accretion will be reduced and/or oxidation of organic soils will be enhanced in the extensive floodplains of the St. Johns River as a consequence of changes in river stage induced by additional water withdrawals. The Phase I study involved calculations based on literature values for release rates of various constituents from flooded and exposed organic soils. None of the literature values of substance release rates used in the Phase I study appears to be from soils in the St. Johns River Basin.

Three of the potential effects were considered to have potentially high significance: (1) reduced nutrient sequestration, (2) increased release of colored dissolved organic matter (CDOM), and (3) increased production and reduced sequestration of greenhouse gases produced within inundated organic soils. Preliminary results were presented for inorganic phosphorus release only, although the work plan for Phase II indicates that nitrogen will be addressed by undertaking appropriate data collection. The Phase I report did not make a persuasive case that changes in CDOM concentrations or loadings under conditions of additional water withdrawals would have significant ecological or water-quality impacts.

If laboratory experiments are undertaken during Phase II to obtain local nutrient and CDOM release rates from drained soils, the workgroup should use procedures that will yield data reflective of environmental conditions. It is critically important that sufficient samples be analyzed to address the well-known large heterogeneity found in soils. Experimental studies should be done at as large a spatial scale as possible to avoid artifacts caused by trying to extrapolate results from small sample sizes and small containers to the ambient environment.

Finally, it may be premature to conduct extensive laboratory and field experiments to evaluate rates of nutrient and CDOM release from drying soils of riparian wetlands in the St. Johns River Basin. A sequential approach would be more effective, in which additional analyses establish the areal extent of wetland soils that would be dried *to a sufficient extent* and *for a sufficient duration* to enhance oxidation of soil organic matter and subsequent release of nutrients and CDOM. If these studies indicate a high likelihood of effect, then experimental studies (at mesocosm rather than microcosm scales) could be undertaken to measure rates of nutrient and CDOM release under environmental conditions relevant to the proposed water withdrawal scenarios.

Plankton, Nutrients, and TMDLs

The plankton and TMDL workgroup was tasked with identifying and quantifying possible environmental impacts of water withdrawals on plankton communities and existing TMDLs in the lower and middle St. Johns River. Overall, the workgroup did a commendable job summarizing and interpreting archival data from a variety of studies conducted in the middle and lower reaches of the St. Johns River over the past 25 years. However, several critical issues were not considered.

First, additional water withdrawals may increase the likelihood, duration, and areal extent of water column stratification and bottom water hypoxia in the lower St. Johns River under low-flow conditions, which was not discussed in the Phase I report. Second, the Phase I report does not adequately address the type or frequency of additional water quality and biological monitoring data needed to adequately assess the impacts of water withdrawals on TMDLs and plankton. Phase II plans include monitoring only in Lake George. Third, it appears that the District has concluded that phosphorus is the limiting nutrient for algal growth in the freshwater portions of the St. Johns River. The occurrence of substantial blooms of nitrogen-fixing cyanobacteria in Lake George is strong evidence that nitrogen limitation occurs at least at some times of year and in some locations within the river.

A key goal for this workgroup during Phase II is to better estimate nutrient and CDOM loading for segments of the middle and lower St. Johns River under the various water withdrawal scenarios and how this will affect plankton dynamics. This will require tighter integration with the hydrodynamics and hydrologic and the biogeochemistry workgroups. The effects of high concentrations of CDOM (that may result from water withdrawals) on phytoplankton ecophysiology were not explored during Phase I, although the biogeochemistry workgroup did consider increased concentrations of CDOM. If further analysis from the biogeochemistry workgroup provides support for an effect of CDOM, then the plankton workgroup will need to consider how this may affect plankton growth in the river.

Phase II work plans for the plankton workgroup indicate the District's understanding of the importance of both nitrogen and phosphorus to plankton dynamics. Plots of nitrogen:phosphorus ratios as a function of potential controlling variables, such as discharge rates, residence time, and water age, could be useful for assessing potential impacts, especially as related to nitrogen-fixing cyanobacterial blooms. Finally, the workgroup is encouraged to continue the development and validation of three-dimensional water quality simulation models (e.g., CE-QUAL-ICM) for the major reaches and lakes of the St. Johns River.

Benthos

The benthos workgroup of the WSIS is studying benthic macroinvertebrates in both freshwater and brackish water environments to better understand the ecological consequences of water withdrawals in the St. Johns River. In the Phase I report, the workgroup reviewed past work in the watershed including an analysis of an existing data base, and it briefly described several conceptual models to guide thinking about how to study and understand the effects of water withdrawal on the trophic organization of benthic invertebrates. In addition to a focus on trophic organization, the Phase I report also targets a few invertebrates (e.g., crayfish, apple snail, blue crab, penaeid shrimp) as taxa of special interest that could be used as species-level indicators in the WSIS.

Although the literature review evident from the Phase I report was extensive, it was not clear what lessons the workgroup would be applying to the St. Johns situation. For example, it was not clear which papers documented how invertebrate assemblages changed as a result of water withdrawal. Another subject not touched in the Phase I report is the effect of water withdrawal on macroinvertebrates with meroplanktonic larval stages, a group of taxa with high relevance in a comprehensive ecological assessment. This subject might be addressed jointly with the plankton workgroup during Phase II.

The need to collect more data was recognized in Phase I, and a first-level plan was fleshed out in the Phase II work plan. In contrast to freshwater areas, the Phase II work plan concludes that no new data will be collected in the estuarine segments of the watershed. Instead, data from existing studies will be analyzed further with the goal of evaluating the effect of salinity and other water-quality variables on invertebrate community structure. Without more information on the context of those historical data collections (e.g., timing, spatial distribution, duration, kinds of data, data collection protocols), it is impossible to judge the merit of that decision. Rather, it would be better to defer any decision about additional data collection from the estuary until a careful evaluation of the utility of existing data is completed. Other factors—such as new dredging by the port authority or the U.S. Navy—that might affect salinity, sedimentation, or other conditions in the lower river may also require reevaluation of estuarine monitoring decisions.

Overall, the committee supports the District's commitment to study the effects of water withdrawal on benthic invertebrate communities. Unfortunately, the available data are insufficient to define precisely what the effects of water withdrawals on the benthos of the St. Johns River will be. The District is commended for sketching the rudiments of a sampling and data analysis program for Phase II, although much of the detail of those efforts is yet to be defined.

Littoral Zone

SAV is a focus of the WSIS because the proposed surface water withdrawals are likely to exacerbate salinity intrusions in the estuarine portion of the river, which could have detrimental effects on local SAV populations. The littoral zone workgroup is assessing the potential damage to SAV from the proposed water supply withdrawals via an extensive monitoring program. By assessing the condition of the SAV during high-salinity pulses in the lower river and comparing these data to literature values on salinity tolerance, the workgroup has made preliminary estimates of the effects of water supply withdrawals on SAV.

The focus of the Phase I effort was the lower 131 km of the St. Johns estuary, where SAV abundance has been monitored intensively over the last 10 years. Most of the cover is due to *Vallisneria americana*, a perennial species with well-developed underground roots and rhizomes that are ideal for consolidating bottom sediments, providing oxygen to benthos and promoting nitrification. In addition, *V. americana* beds are excellent habitats for small fish. During Phase I, the District predicted that projected future water withdrawals could have dramatic consequences on SAV in some areas, especially where *V. americana* populations now fluctuate in the lower St. Johns River. Although *V. americana* presumably could migrate further upstream, there is less shallow water area there, so a net loss of habitat is still expected.

Improved hydrodynamic and hydrologic modeling during Phase II is expected to provide more spatially explicit predictions of the salinity increases in the littoral zone. To enhance their monitoring program, the District should consider adding at least one continuous salinity monitoring station in the littoral zone during Phase II to detect short-term salinity excursions where *V. americana* is at risk. The workgroup should also undertake more study of salinity tolerance of local populations from the St. Johns River, perhaps via mesocosm studies, in order to validate the values derived from the literature. Finally, the workgroup might assess whether any other existing SAV species, for example *Ruppia maritima*, might be able to take the place of

V. americana as a dominant macrophyte in the littoral zone. A mesocosm program similar to the one described above for *V. americana* would be helpful in this regard, although it is acknowledged that such experiments go beyond the planned Phase II work.

Fish

The District's Phase I report describes how water withdrawals could influence spawning success and recruitment of important recreational and commercial fishes, populations and distribution of other fish species, and critical dimensions of fish habitat. Two potential effects were considered by the fish workgroup: (1) the direct effects from entrainment or impingement and (2) indirect effects associated with changes in habitat caused by water withdrawals.

The District's hypothesis is that given the relatively low intake velocities of the proposed water withdrawals, adult and juvenile fish entrainment or impingement probably will not be significant enough to elicit broad-based community changes. However, given the lack of information on entrainment and impingement of larval stages, the District has begun a study of larval fishes in five regional locations in the upper and middle basins of the river. It will be important for the District to clearly define the actual distribution and frequency of Phase II sampling, as this was not evident in the Phase I report or the Phase II work plans. The committee's concerns include the frequency of larval fish sampling in these areas, the lack of nocturnal sampling, and the narrow focus on clupeid larvae.

The Phase I report describes other potential environmental impacts to fishes as a result of water withdrawal, which can be much more difficult to assess. With respect to salinity effects, the District had contracted with the Florida Fish and Wildlife Conservation Commission to examine the influences of water level and salinity changes on fish in the lower river. Although this is a step in the right direction, there are significant limitations to this monitoring that need to be kept in mind. For example, this program does not collect samples in small creeks in the lower river and thus will miss important nursery areas. Finally, the fish data analyses need to be integrated with other components of the regional biota (e.g., benthos, decapods, SAV) that also may be influenced by salinity shifts.

One last concern with the Phase I report is the lack of consideration of how water level decreases will influence fish population dynamics and distribution. The District should carefully study potential impacts to all fishes in the middle and upper St. John River but particularly those that require shallow areas for spawning and foraging, including but not limited to the centrarchids. The Phase II work plan makes progress in explaining the District's approach to examining how water level changes might impact floodplain habitat for fishes.

Wetlands and Wetland-Dependent Species

Changes in hydrology can alter the structure and function of wetlands. Therefore, the wetlands workgroup is examining the potential impacts of the proposed surface water withdrawals to wetland vegetation communities and a few species of wetland-dependent fauna. Impacts range from changes in vegetation community type or structure (including species composition) and altered productivity to shifts in the position of boundaries between communities. The Phase I wetlands work was descriptive and conceptual in nature.

A major concern of the committee is how the workgroup identified and delineated wetlands that are influenced by the river's hydrology. The floodplain was delineated using the 5-foot contour line, with the assumption that the 5-foot elevation captures much of the 50-year floodplain. Unfortunately, this crude approach lacks the resolution necessary to predict the ecological effects of hydrologic change. Fine-scale elevation data in the form of a digital elevation model are needed to produce accurate maps in flat, wetland-rich areas (particularly in efforts to characterize hydrology); map wetland diversity; and anticipate likely changes in these distributions as a result of water withdrawal. The District plans during Phase II to produce digital elevation models for the portions of the watershed where LIDAR data are available.

The wetlands workgroup divided the river into nine segments deemed relatively homogeneous in terms of soils, vegetation, hydrology, water quality, and fauna. For each segment, qualitative information was provided on the relative likelihood of impacts from water withdrawals to wetlands. The criteria used to determine the likelihood of impacts are not provided in the report, making it virtually impossible to judge the results.

Finally, the District should consider broadening the range of taxonomic groups used in monitoring wetland impacts. Many wetlands in the floodplain are not used as habitat by the four chosen avian species, a problem compounded if the bird populations occur in low densities. Amphibians, reptiles, invertebrates, and plants have proven to be valuable indicators of hydrologic impacts to wetlands in monitoring programs and might be considered as indicators in Phase II. The district has proposed investigating additional species in the Phase II work plans; however, there is no detail or methodology provided on how the different assemblages presented (e.g., reptiles, amphibians) might be used to indicate impacts.

1

Introduction

The St. Johns River in central and northeast Florida is a 310-mile-long waterway that flows north from Indian River County to the Atlantic Ocean. The river and its floodplain comprise a broad, low-gradient system that transitions from dominance by freshwater wetlands in the headwaters near Vero Beach to dominance by a wide estuarine channel in its lower reaches near Jacksonville. The river is the longest in Florida, and drops only 30 feet from its headwaters to its mouth at the Atlantic Ocean. The lower 100 miles are subject to tidal influence.

The St. Johns River Water Management District (SJRWMD or "the District"; see Figure 1-1) is responsible for managing water resources in the basin and its surroundings—an area of 12,283 square miles, or 23 percent of Florida. The District must provide for water to meet the needs of human society and protect natural systems within the river basin. Each task is challenging in its own right, but together they require careful integration of many technological, scientific, social, economic, and environmental factors. There are approximately 4.4 million people (21 percent of Florida's population) living in the area served by the District and 110 municipalities, including the growing cities of Jacksonville, Orlando, and Gainesville. The District expects an increase in basin population from 3.5 million in 1995 to 7.2 million in 2030 (SJRWMD, 2009a).

In order to meet the water supply needs of the District's residents and its industrial and agricultural users (the total demand was 1.49 billion gallons per day in 2000), the District historically has relied on groundwater, with the upper Floridan aquifer as the primary source for public supply (SJRWMD, 2006). As a result of its ongoing water supply planning efforts, however, the District has determined that its useable groundwater supplies will reach their sustainable limits in the near future, prompting a search for alternative water supplies. Indeed, groundwater withdrawals in the Central Florida Coordination Area (CFCA) that lies in the southwest portion of the District, near Orlando, are expected to reach their sustainable limits in the year 2013 (SJRWMD, 2006). The predicted growth in withdrawals from the Floridan aquifer for most of the rest of the District is not sustainable through the 2030 planning horizon (SJRWMD, 2009a). Alternative water supplies under consideration include four new surface water withdrawal sites in the St. Johns River, one in the lower Ocklawaha River, and a desalination facility on the Coquina Coast (see Figure 1-2).

Due partly to stakeholder concern about the shift to reliance on surface water sources, the District has embarked on an intensive study of the potential impacts of increased surface water withdrawals in the St. Johns River basin (described in more detail later in this chapter), and, in 2008, it requested that the National Research Council (NRC) form a committee to review scientific aspects of the impact assessment and provide advice to its ongoing effort. To set the context of this report, this introduction begins with a discussion of the District's water resources

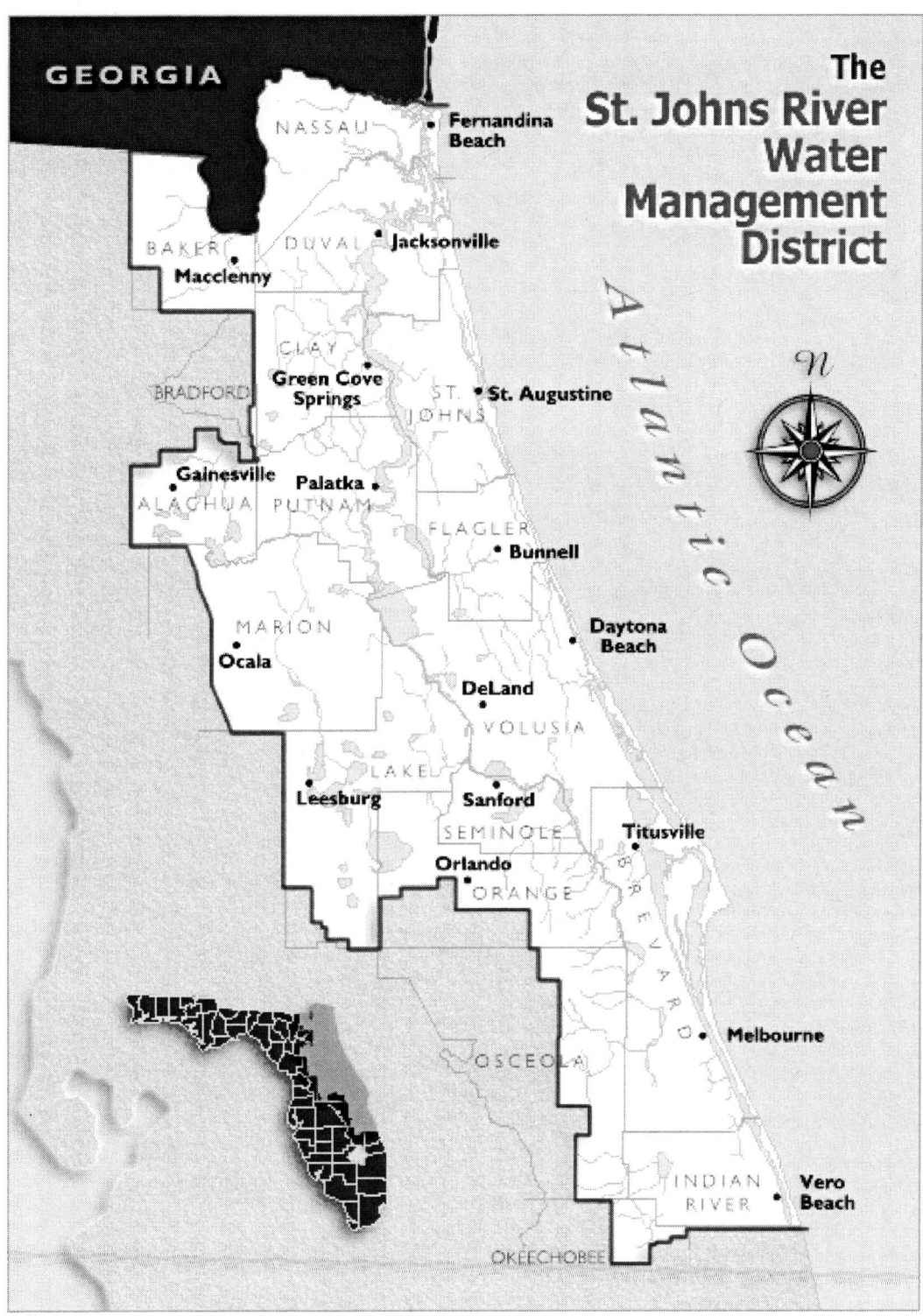

FIGURE 1-1: The St. Johns River Water Management District.
SOURCE: http://sjrwmd.com/about.html.

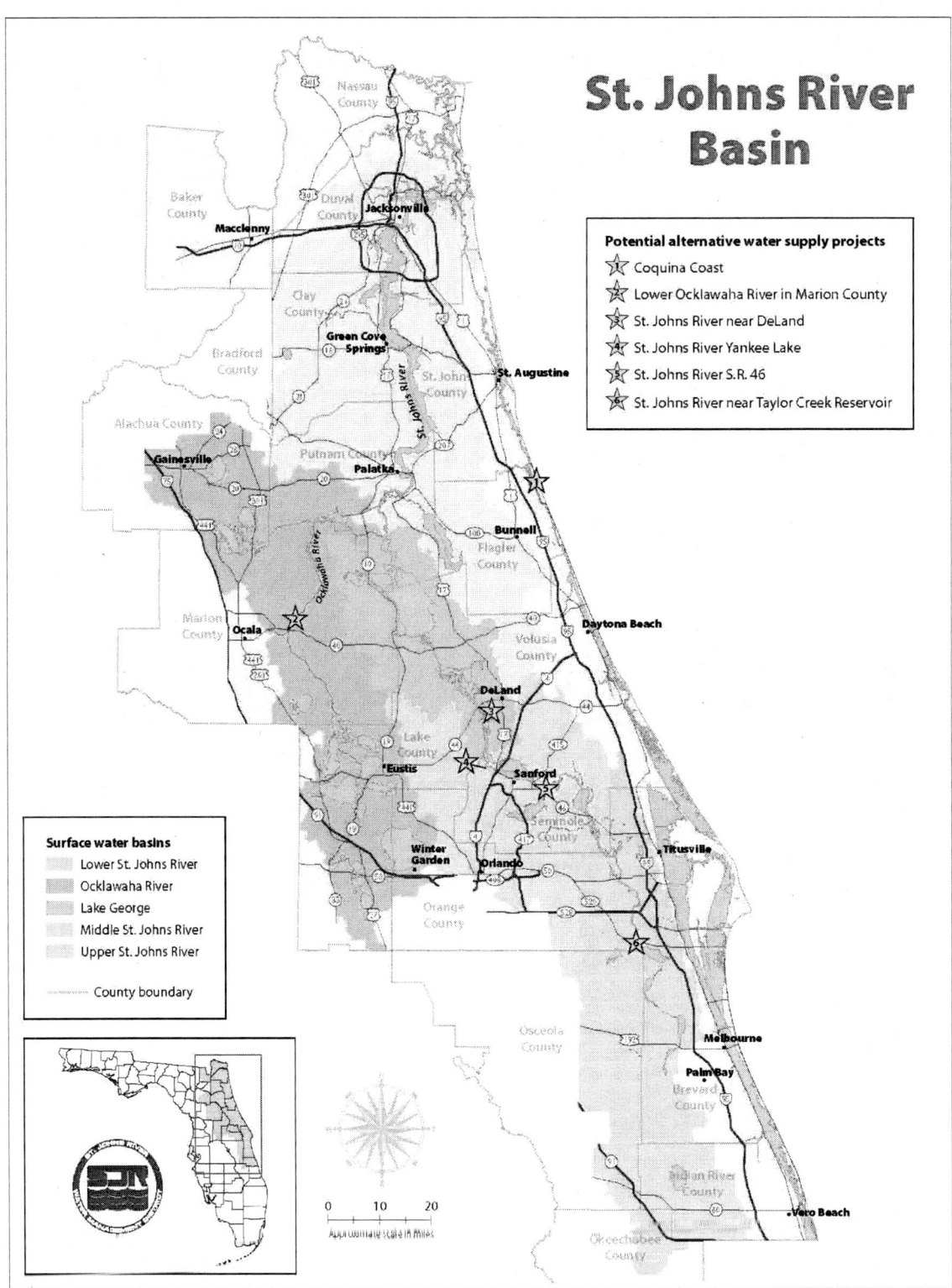

FIGURE 1-2: The five surface water basins of the St. Johns River Water Management Basin and the location of six alternative water supply projects, including five surface water withdrawal sites.
SOURCE: Tom Bartol, SJRWMD, personal communication, 2009.

Introduction

planning process, including how it identifies future water demands, evaluates water resources, and selects alternative water supply options. The chapter then describes the District's Water Supply Impact Study and the role of the NRC.

SJRWMD WATER RESOURCES PLANNING

The District is tasked with (1) ensuring that adequate water supplies are available to meet future demands, and (2) preventing the occurrence of unacceptable impacts to water resources and related natural systems. As population growth and associated land development continue, these goals can increasingly come into conflict with each other. Thus, careful water supply planning is required.

Water Resources Assessments

The District works closely with water suppliers to ensure that adequate water supplies are available to sustainably address future demands, and it forecasts water demands for public and private needs, including domestic, agricultural, commercial, industrial, and power generation purposes. The most recent water demand forecasts, developed collaboratively with the District and area water suppliers, are outlined in the 2008 Water Supply Assessment (WSA; SJRWMD, 2009a). The latest assessment states that total average annual demand is expected to increase from 1,346 million gallons per day (MGD) in 1995 to 1,742 MGD in 2030.[1] The WSA is produced on a five-year cycle, allowing for the successive application of improved analytical tools and resulting refinement of long-term plans (e.g., incorporating potential effects of climate change). Such frequent water demand forecast production cycles have been shown to be beneficial (Hagen et al., 2005). The demand forecasting process for the 2008 WSA used a planning horizon of the year 2030. A horizon of approximately 20 years is typical and appropriate for water management planning, especially considering the unavoidable uncertainties in the supporting forecasts and the District's five-year cycle of updating. The District also makes projections to the year 2050 to identify longer-term trends and raise awareness of potential future demands.

Based on its demand projections, the District analyzed the adequacy of available water resources, including consideration of conservation, water reuse, and the development of new sources. The District's approach to water supply assessment assumes that the 2030 planning horizon year is one of average rainfall. However, the assessment is also conducted assuming 1-in-10 year drought conditions, which provides some degree of sensitivity analysis. From the demand and resource analyses, the District identified approximately 40 percent of its area as Priority Water Resource Caution Areas (PWRCAs), where anticipated sources of water and conservation efforts would be inadequate to meet future demands while sustaining water resources and related natural systems through 2030. Further preliminary analysis has led to almost all of the remainder of the District being identified as "potential" PWRCAs. Therefore,

---

[1] Data for 2005 is considered atypical because of high rainfall and low demand.

the District has designated its entire 18-county jurisdictional area as one water supply planning region pursuant to the requirements of Subparagraph 373.036(2)(b)2 of the Florida Statutes.

There are several alternatives available to the District to meet water future needs, including conservation, reclaimed wastewater, brackish groundwater, surface water, and seawater. All are being used today to some extent, and the District imposes stringent conservation and wastewater reuse provisions in its water appropriation permits. The resulting treated wastewater is currently put to many secondary uses, such that 74 percent of treated municipal wastewater effluent in the SJRWMD is beneficially reused (SJRWMD, 2008). The extent of wastewater reuse varies widely by county, from being reused more than once in Orange, Indian River, Osceola, and Okeechobee counties to only 9 percent reuse in Duval County (FDEP, 2009).

Florida Statute 373.019(1) defines alternative sources, in part, as seawater or brackish surface water or groundwater, captured wet weather flows, stormwater, or other non-traditional sources as designated in the applicable water supply plan. In its most recent Water Supply Plan (SJRWMD, 2006), the District presented many possible water supply development project options from which water utilities could select to address future supply needs including new or expanded brackish groundwater desalination facilities, surface water withdrawal sites, and seawater desalination plants.

Before the District issues permits for new consumptive use withdrawals from a water source, it determines how much water is available (if any) for human use without harming the resource and its related natural systems. This is accomplished via the derivation of a minimum flow and/or level (MFL) for the relevant surface water or aquifer source. An MFL (particularly for a surface water source) takes into account seasonality and varying flow regimes in order to protect biological communities. At a given location, the MFL describes the flow magnitude, duration, and frequency necessary to prevent harm to ecological communities under a variety of high-, average-, and low-flow conditions. The MFL for the St. Johns River near DeLand is based upon an analysis by Mace (2006). Water withdrawal scenarios that do not violate MFLs can be developed in infinite variations of seasonal pumping rates and short-term storage volumes. Of the many possible scenarios, a withdrawal rate of 155 MGD was determined to be the maximum continuous yield at DeLand that would meet all MFL criteria without requiring storage (see Robison, 2004). The MFL for the lower Ocklawaha River basin is expected to be developed in the near future. MFLs are the result of statistical analyses of available and historic data and are developed to be protective of the water and ecological resource. The District conducts periodic reassessment of adopted MFLs based on consideration of new data and information as they become available (SJRWMD, 2006). Consumptive use permits are typically issued for periods of 20 years, but they may be re-opened at any time, and restrictions may be imposed on one or more use categories in response to reassessments of source yield and/or water shortages.

Selection of Potential Surface Water Withdrawal Sites

In order to determine potential sites for public water supply withdrawal along the St. Johns River, the District employed a structured process of data assessment, feasibility evaluation, potential project selection, and quantitative water supply availability and yield analysis (CH2M Hill, 1996a, 1996b, 1997). Briefly, this process requires first evaluating data from long-term river gauges. Yields at these gauge locations are then screened on the basis of magnitude, variability, minimum instream requirements, and other withdrawal constraints. The yield information is compared spatially with expected growth in demand at population centers in order to balance meeting demands with the least distance from a withdrawal point of adequate yield, which may or may not be coincident with a flow gauge (SJRWMD, 2006). This quantitative analysis led to the selection of the sites shown in Figure 1-2, which could supply a total combined yield of 155 MGD from the headwaters to DeLand. Additionally, in-stream monitoring and treatability studies at these sites indicated that several effective and efficient water treatment combinations (including desalting) could be used to produce water of a quality that is suitable for use as public supply and at an affordable cost (CH2M Hill, 2004; Burton and Associates, 2004).

The lower Ocklawaha River basin was identified as a potential source in the District's 2000 Water Supply Plan (SJRWMD, 2000) and was the subject of a separate analysis by Hall (2005), in which it was estimated to have a relatively large potential water supply yield of 107 MGD. However, it was acknowledged that a more accurate assessment of water availability would only be determined with the development of an MFL. Unique hydrologic factors make this a favorable resource for surface water supply development because inflow to the Ocklawaha River includes the 876 MGD discharge from Silver Springs. The water quality of the lower Ocklawaha River is very good, due in large part to this substantial fresh groundwater contribution. The combination of relatively good raw water quality and significant base flow make this an attractive candidate for surface water supply development. Many other smaller resource development projects have been identified as possible options for providing minor amounts of supply (SJRWMD, 2009b).

WATER SUPPLY IMPACT STUDY

As described in the previous section, recent water supply planning efforts, including the development of statutory minimum flows and levels (MFLs) for portions of the St. Johns River, indicate that up to 262 MGD could be withdrawn from the St. Johns (155 MGD) and Ocklawaha (107 MGD) rivers for public water supply without causing significant environmental harm. To more fully explore this possibility and at the request of the District's Governing Board, in January 2008 the District embarked on a two-year, two-phase study called the St. Johns River Alternative Water Supply Cumulative Impact Assessment (CIA).

The District initiated the CIA to (1) evaluate the potential environmental effects of withdrawing water from these rivers, and (2) alleviate public concern by ensuring that the best possible scientific effort has been made to evaluate potential environmental effects. Due to confusion with respect to its title, the District has subsequently renamed the study the "St. Johns River Water Supply Impact Study" (WSIS) to more accurately reflect its goals. In order to maintain consistency with future NRC reports on this topic, the District study will be referred to as the WSIS in this report.

The WSIS is being conducted by District staff with the assistance of several outside experts and is divided into seven workgroups:

1) Hydrologic and hydrodynamic modeling,
2) Biogeochemistry,
3) Plankton, nutrients, and total maximum daily loads,

4) Benthos,
5) Littoral zone/submersed aquatic vegetation,
6) Fish, and
7) Wetlands and wetland-dependent species.

Each workgroup is examining its defined issues across the entire St. Johns River system, from the headwaters to the mouth, from the channel to beyond the margin of the floodplain, and from the soil and channel substrate to the water surface and above. Impacts to the Ocklawaha River basin are not being considered as part of this assessment because they will be addressed during the development of the MFL. However, impacts on the lower St. Johns River from withdrawal of 107 MGD from the Ocklawaha River will be included.

In Phase I of the WSIS, existing models and available data were used to examine potential changes in hydrology caused by withdrawals and consequential impacts on salinity, wetland soil chemistry, flora and fauna of the river channels and associated wetlands, and effects on the littoral zone. The results of the hydrologic analyses consist largely of ranges of responses of river and wetland water flow, velocity, and depth, and the upstream extent of salinity. The biochemical analyses conducted in Phase I posited a wide range of potential impacts due to the alteration of hydrologic conditions under the influence of increased withdrawals from the river. Quantitative ranges of the biochemical impacts were developed from reaction rates extracted in the literature, and applied over the areal extent of susceptible riverine and wetland regimes. Analytical and data deficiencies will be reduced in Phase II by focused model improvement and the collection of needed monitoring data. Specific environmental impacts are expected to be better determined in Phase II. The District will use this information to delineate environmental effects boundaries and assess the potential for crossing response thresholds at various levels of water withdrawal. In each case, in order to assess the significance of these effects they will need to understand (1) the strength of the effect, (2) its likelihood, and (3) the persistence of that effect relative to recovery time. In Phase II the District also plans to recommend ways to avoid or minimize adverse hydrologic effects (e.g., low-flow cessation, altered timing of withdrawals, and design of intake structures).

NRC STUDY AND REPORT ROADMAP

The District has requested that the National Academies provide peer review and advice as the study progresses. The task of this NRC Committee is to review and critique the ongoing work of the WSIS via short reports and a final report (see Box 1-1).

This first report is based upon a review of the draft report of the Phase I work (SJRWMD, 2008), an onsite meeting with District staff and stakeholders in January 2009, and information subsequently requested and received by the committee including a document describing plans for Phase II studies by the workgroups (SJRWMD, 2009c). As is often the case for large projects that are developed rapidly by multiple research teams and authors, the sections of the Phase I report can be criticized readily for many deficiencies in presentation, missing data, and incomplete analysis. However, the Phase I report is a first step in a complex endeavor that crosses many disciplines. This NRC report does not include critiques of details that are inherent in the preliminary nature of the Phase I study. Instead the report focuses on the key points that will lead to necessary improvements in the Phase II study. Furthermore, because this report is the first in a series, it does not attempt to cover every possible issue related to water withdrawal

> **BOX 1-1 NRC STATEMENT OF TASK**
>
> An NRC committee overseen by the Water Science and Technology Board of the National Academies will review the progress of the St. Johns River Alternative Water Supply Cumulative Impact Assessment, now known as the Water Supply Impact Study (WSIS). Communities in the St. Johns River watershed in east central Florida are facing future drinking water supply shortages that have prompted the St. Johns River Water Management District (the District) to evaluate the feasibility of surface water withdrawals. At the current time, drinking water is almost exclusively supplied by withdrawals from groundwater. Reliance on groundwater to meet growing the growing need for public supplies is not sustainable. The St. Johns River and the lower Ocklawaha River are being considered as possible alternatives to deliver up to 262 million gallons of water per day to utilities for public supply. In January 2008, the District began an extensive scientific study to determine the feasibility of using the rivers for water supply, and it has requested the advice of the National Academies as the study progresses.
>
> The WSIS is composed of six major tasks, being carried out by District staff scientists aided by a suite of outside experts, each with national standing in their scientific discipline. These activities include modeling of the relevant river basins, determining what criteria should be used to evaluate the environmental impacts of water withdrawals, evaluating the extent of those impacts, coordinating with other ongoing projects, and issuing a final report. The NRC committee will review scientific aspects of the WSIS, including hydrologic and water quality modeling, how river withdrawals for drinking water will affect minimum flows and levels in the two rivers, the impact of removing old and introducing new wastewater streams into the rivers, the cumulative impacts of water withdrawals on several critical biological targets, and the effects of sea level rise. Potential environmental impacts being considered by the District include altered hydrologic regimes in the river, increased pollutant concentrations in the rivers (e.g., sediment, salinity, nutrients, and temperature), associated habitat degradation, and other direct effects on aquatic species due to the operation of the new water supply facilities.

from the St. Johns River, but rather focuses specifically on the District's Phase I report. Thus, longer-term subjects such as changing land use in the St. Johns river basin, post-WSIS monitoring of the resource to accompany ongoing water withdrawals from the St. Johns, and future climate change are not addressed in this report, but will be considered for inclusion in subsequent reports of the committee. Finally, the report does not discuss predisturbance conditions in the basin because they are not used as the baseline conditions in the WSIS.

This report is organized along the lines of the seven workgroups. Chapter 2 presents several overarching issues and conclusions of the committee about the Phase I Draft Report. Tackling these issues should be the highest priority of the District as it commences Phase II of the WSIS. Chapter 3 focuses on the extensive hydrologic and hydrodynamic modeling effort underway that supports the remaining biogeochemical and ecological analyses, and Chapter 4 reviews the Phase I work of the biogeochemistry, plankton, benthos, littoral zone, fish, and wetlands and wetland-dependent species workgroups. Whenever possible, conclusions and recommendations made for one workgroup that are critical or relevant to another workgroup are mentioned in both places.

2

St. Johns River Water Supply Impact Study: Cross-Cutting Issues

The St. Johns River Water Management District (SJRWMD or "the District") has undertaken a large and complicated study to evaluate potential environmental impacts of additional consumptive-use water withdrawals from the St. Johns River and its main tributary, the Ocklawaha River. The study involves nearly one hundred scientists and engineers divided into seven workgroups addressing topics ranging from the effects of withdrawals on river hydrodynamics and hydrology to water quality and a variety of biological and ecological issues. The Phase I report of the study (SJRWMD, 2008) covers initial analyses by the workgroups on all of these topics and totals more than 700 pages of text, tables, and figures.

The District is commended for the substantial financial and human resources that they have given this important study, and the workgroups also are commended for the large amount of work they accomplished in a short period of time. Phase I of the study identified important issues related to the proposed water withdrawals, established many of the kinds of information needed to assess impacts of the withdrawals, and for many of the topics made an initial evaluation of the availability of this information. The Phase I report is not very instructive, however, regarding plans for the scope of Phase II studies and how they will be conducted. Based on a request from the committee at its first meeting in January 2009, District staff produced a document describing methods they intend to use in Phase II of the study (SJRWMD, 2009c). That document, hereafter referred to as the Phase II work plans, was provided to the committee prior to its second meeting in May 2009, and it was used by the committee to inform its analysis of the Phase I study and its recommendations for work to be undertaken in Phase II.

This chapter is focused on the "big picture"—the committee's assessment of the St. Johns River Water Supply Impact Study (WSIS) as an integrated whole—and on issues that cut across or transcend the discipline-based topics of the individual workgroups. General issues such as overall study focus, goals, and design; integration of information across workgroups; and the importance of conceptual models, as well as quantitative simulation models, are the primary focus of this chapter. However, more specific issues such as the importance of quantifying return flows, role of ecological indicators, and assessment of impacts from floodplain dewatering also are discussed, along with the committee's recommendations for dealing with them in the WSIS. The committee feels that these should be the highest priority issues for the District to consider as Phase II commences.

More specific assessments of the studies conducted by the seven workgroups are given in Chapters 3 and 4. The former chapter deals with studies by the hydrology workgroup on potential changes in the river's hydrodynamics and hydrology resulting from the withdrawal of an additional 262 million gallons per day (MGD) of water from the main stem of the St. Johns River and the Ocklawaha River. Chapter 4 addresses studies conducted by the other workgroups, which collectively are responsible for addressing the potential water quality and ecological impacts of changes in river hydrodynamics and hydrology.

STUDY FOCUS

The committee notes that ecological impacts along the largest tributary to the St. Johns, the Ocklawaha River, were not being assessed to the same degree in the WSIS as in the St. Johns River itself. Increases in water withdrawals in the system have the potential to affect hydrologic and ecological conditions in the Ocklawaha, as well as the main channel of the St. Johns, and this is particularly likely given that some of the proposed additional withdrawals will occur from the area near the confluence of the Ocklawaha with the St. Johns River. The committee understands that exclusion of the Ocklawaha from the current WSIS on the St. Johns River was a conscious decision and not an oversight. The decision was based on three reasons. First, the Ocklawaha is hydrologically distinct from the St. Johns, and somewhat different issues likely would be important in an analysis of that system. Second, for financial and human resource reasons SJRWMD managers concluded that a focused study on the St. Johns River itself was more feasible and that including the Ocklawaha would dilute the effort needed for a comprehensive analysis on the St. Johns. Third and probably of greatest importance, a separate, formal (i.e., legally based) analysis of minimum flow and level (MFL) requirements for the Ocklawaha River is planned for the near future (date not yet specified by the SJRWMD). The MFL process would not be as elaborate as the current WSIS, but it would represent an additional several hundred thousand dollar effort on the Ocklawaha River.

The committee does not dispute the reasoning of the District in deciding to focus on the St. Johns River. Hydrologic changes in the Ocklawaha River in fact are not being excluded from the St. Johns River WSIS but are being treated as "input conditions" for the assessment of impacts along the St. Johns River. Given the near-term plans (not yet fully specified) for a separate MFL analysis on the Ocklawaha River, it seems prudent to focus the current analysis on the main-stem St. Johns. Nonetheless, the committee recognizes the need for comparable impact analyses on the Ocklawaha River, particularly its lower reaches, which are the most likely to be affected by additional water withdrawals in the basin. In addition, the committee concludes that to avoid future misunderstandings the SJRWMD should provide a better explanation of the basis for its decisions on this issue in public documents, presentations, and web-based materials related to the WSIS and should make the focus of the WSIS on the main stem St. Johns River clear in its publications and presentations.

Another concern about the study focus is the lack of attention in the WSIS to the impacts of growth in the human population of the drainage basin that the appropriation of an additional 262 MGD of water will enable. A widely used rule of thumb in water supply planning states that per capita use of water in municipalities on average is about 100-150 gallons per day. This includes water used in the commercial sector, for fire fighting and public use (e.g., in parks), as well as consumption within individual homes. A total additional withdrawal of 262 MGD from the St. Johns River thus would enable approximately 1.7 to 2.6 million more people to live in the drainage basin (assuming no reuse of the withdrawn water occurs). That population, depending on where it is located, could have small or major impacts on water quality in the St. Johns River and its tributaries. The committee recognizes that population growth is a sensitive issue in Florida, that the District has no direct control over this issue, and that accurate predictions of impacts resulting from the additional population cannot be made without more knowledge about the geographic locations where the growth would occur. Nonetheless, the committee recommends that the District at least acknowledge the potential impacts of this growth in its assessment documents, even if it cannot quantify the impacts.

NEED FOR INTEGRATION

Almost any study of the magnitude of the St. Johns River WSIS must be subdivided into components that can be addressed by individuals or small work groups in order to be accomplished in a reasonable timeframe. The committee thus understands and supports the District's approach to divide the WSIS among seven workgroups. The danger in such an approach, however, is that the study could become seven separate studies each with narrowly focused conclusions in the subject expertise of the workgroup rather than an integrated assessment in which output from one workgroup serves as input to another group for an iterative analysis of ecosystem impacts. For example, changes in the benthic macroinvertebrate assemblage may engender changes in the fish assemblage, which in turn may stimulate changes in other segments of the biological community, and feedbacks also occur, making cross-workgroup collaborations essential to a comprehensive analysis.

In oral discussions with staff from the District, the committee has been assured that the project managers and staff agree on the need for integration and are experienced at doing this. However, the written document available for the committee to review, the Phase I report, does not provide much evidence for integrative and cross-workgroup analyses. In part this may reflect the short timeframe allowed for Phase I work to be accomplished and the report to be written. The committee strongly recommends that this situation not occur again in Phase II. Sufficient time and project funds need to be allotted to integrative studies and to cross-workgroup analyses before the final report is written. The committee notes that integration is not an "automatic outcome" in any large, multidisciplinary undertaking; it needs to be planned for and managed.

Applying a Conceptual Framework to Integrate Studies

During two of the District's presentations to the committee, conceptual frameworks depicting how various parts of the river system interact were shown that did not appear in the project's written reports. **As a way of integrating the work of the seven science groups, it is recommended that the District further develop these conceptual frameworks of qualitative interactions.** The term "framework" is used here instead of "model" to clearly distinguish between an overarching organizational framework and the discipline-specific models used in different parts of the study. This framework should provide a means for linking the various ecological and physical relationships affected by surface water withdrawals. The framework should be built from the bottom-up, with the validity and importance of each linkage being independently evaluated. Each process and linkage should be identified as to whether it is demonstrated by data or by model predictions, or whether it is a hypothesis or assumption, thereby allowing cross-disciplinary understanding of the limits of existing science for the St. Johns River. The framework should *not* be simply a set of one-way descriptive boxes; that is, if it merely illustrates the subsidiary processes and linkages, it will be of limited use. Instead, the District should focus on a framework with feedbacks from individual research project goals and methods that are specific to the St. Johns River. Each piece of the research program should be motivated to fill in specified unknowns in the conceptual framework boxes and linkages.

An example of a complete and comprehensive conceptual framework is the one developed for the Florida Everglades (http://www.evergladesplan.org/pm/recover/cems.aspx). It is not expected that the District can develop a similarly comprehensive set of concepts within the scope of the present project. Furthermore, simple adaptation of the Everglades framework is discouraged. However, a more modest conceptual framework might be represented as either a matrix or flow chart that links possible physical effects caused by surface water withdrawal to the perceived chain of ecological effects and their consequences and the resulting research needs. This approach can be used to better connect the field studies, models, and analyses. Such a framework should explain (1) how the field data collected are appropriate to the model/analysis needs, and (2) how the model/analysis outputs will answer the key questions to quantify the linkages and processes that are unknown. The District should use this framework to help formulate which studies form the critical paths to quantifying impacts. The overall goal is to develop a more holistic and linkage-based understanding of the interplay between the different ecological disciplines and the physical process, while ensuring that critical pieces are not neglected. In the spirit of adaptive management, the framework should be revised and updated as new data and information are collected, analyzed, and modeled, and it should provide feedback to help modify monitoring and research strategies (as discussed below). Development of such a framework will be important for scientists within the District to better communicate with one another as well as with their consultants, this committee, and other reviewers of their work including the public.

Each of the workgroups of the WSIS is using models to conduct scenario analyses. The District staff should evaluate what should and can be done to connect the models among the workgroups. As a minimum, the assumptions used by one workgroup in running a particular model need to be apparent to other workgroups who intend either to use the model directly or to use model output produced by the first workgroup in their own analyses. The committee is aware of the difficulties involved in linking models that have substantially different computational time-steps and different spatial scales, but efforts to link the models, where feasible, would doubtless lead to more sophisticated and comprehensive analyses, and the committee encourages the District staff to move in that direction.

Finally, in both modeling and synthesis of the findings from the WSIS, the District staff and their consultants are encouraged to incorporate uncertainty analyses to the extent possible. Reliance on mean values produced by models or other analyses is highly unlikely to reflect the richness of information obtained by the study and also unlikely to reflect the range of impacts of additional water withdrawals on the St. Johns River and associated ecosystems.

Using Testable Hypotheses and Quantitative Research Questions

Integration would be more easily achieved if the District were to provide a clearer set of testable hypotheses and quantitative research questions that link the hydrodynamics and hydrology to the ecology and reflect the state of knowledge along with the planned studies. That is, the proposed conceptual framework described above provides organizing principles for qualitative understanding of the linkages, whereas the hypotheses and research questions provide the guidelines for the scientific efforts required to quantify the linkages. As an example, an overarching set of physical hypotheses might be framed as follows: withdrawal of surface water from the middle St. Johns River will result in one or more physical consequences: (1) increased reverse flow within the lower St. Johns River, (2) increased downstream flow in the upper St. Johns River, (3) increased groundwater discharge into the river, and (4) reduced water levels. These physical consequences are sketched in Figure 2-1.

All the possible consequences should be paths in the conceptual framework, but an organized plan of testing is required to identify and quantify key processes and determine which may be omitted from further consideration. To continue this example, subsidiary physical hypotheses might be: (a) increasing reverse (upstream) flow leads to increasing salinity in the lower St. Johns River due to tidal transport of oceanic water; (b) increased downstream flow in the upper St. Johns River leads to higher velocities and a reduction in wetland area, and (c) increased groundwater flow leads to a lowering of the water table and/or increased saltwater discharge to the river. Each of these physical hypotheses could be linked to hypotheses associated with ecological behavior (for example, increase salinity in the lower St. John River could have negative effects on beds of submersed aquatic vegetation). A similar set of hypotheses could be generated for the effects of sea-level rise. With these hypotheses as the research guidelines, the District could ensure that the individual research projects will provide answers to research questions framed around quantifying the individual paths and processes.

Creating hypotheses and research questions as part of a conceptual framework can build on the prior studies of modeling on the lower and middle St. Johns River. The modeling efforts were based on model availability and an initial analysis of what constitutes the major issues. From the Phase I screening studies it appears that there are two likely physical impacts of the proposed surface water withdrawals: (1) increased upstream propagation of salinity in the lower St. Johns River caused by reduced downstream flows below the surface water withdrawal, and (2) alteration of the flow and velocity regimes in the middle St. Johns River that may lead to changes in velocity regimes and possibly increased water age (i.e., the length of time a water parcel remains in the river). The Phase I modeling studies did not consider the upper St. Johns

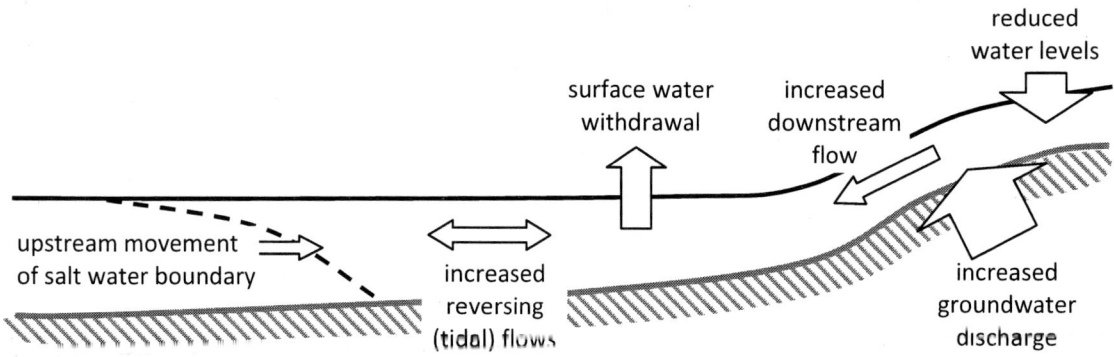

FIGURE 2-1: Sketch of hypothesized chain of physical consequences caused by surface water withdrawals.

River and possible effects of increased groundwater discharge, leaving substantial unanswered questions as to whether the surface water withdrawal will be compensated for by increased groundwater discharge, drying of wetlands in the middle and upper St. Johns River, or reversed flow from the lower St. Johns River. These issues should become part of the conceptual framework created during Phase II of the WSIS.

Investigating Individual Species as Indicators of Community Change

Different workgroups of the WSIS are analyzing potential impacts of additional water withdrawals on several major types of biota: submerged aquatic vegetation (SAV), other wetland vegetation, benthos, fish, and birds. Committee concerns related to individual types of biota are discussed in Chapter 4, but a cross-cutting concern that transcends the individual cases is the basis by which decisions are made to focus on particular species in the WSIS. It generally is not clear to the committee whether the most sensitive species are being considered. Moreover, reliance on single species as indicators of change may underestimate broader biological consequences and miss some of the clearest and most easily interpreted signals of the effects of water withdrawal. That is, **a focus on individual species may limit the District's ability to detect and understand a variety of important ecological dynamics caused by water withdrawals.**

The past two decades have seen the development of more integrative, quantitative methods to characterize the multifaceted aspects of ecological condition (Karr, 2006; Fore et al., 2007; Yoder and Barbour, 2009; Pont et al., 2009). The most integrative of these methods involve (often multimetric) biological indexes, predictive models, or combinations of both approaches. Both American and European governments have acknowledged the utility and need for these integrative ecological approaches (e.g., EPA, 2003, and Pont et al., 2007, regarding the European Water Framework Directive). Widespread successful application of biological indexes for all the major taxa (vegetation, benthos, fish, and birds) and in the major environment types that dominate the St. Johns River basin (wetlands, freshwater streams, and estuaries) underscores their power in understanding and predicting the effects of water withdrawal on St. Johns ecosystems.

A clear example of the need for integration (that is elaborated on in Chapter 4) is provided by the fish workgroup. Fishes inhabiting a river as large and varying over its length as the St. Johns have complex life histories. Many estuarine-dependent fishes spawn off shore (or near shore); the larvae are recruited into estuaries and settle into nursery habitat where they grow to late juveniles or adults; and later, adults migrate back offshore to spawn (Beck et al., 2001). In contrast, some freshwater fishes move along salinity gradients from fresh water to low salinity segments (Peterson and Ross, 1991; Wagner, 1999, Wagner and Austin, 1999). The overlap of these two faunas makes estuarine systems diverse and productive (Peterson and Meador, 1994). Thus, a full understanding of the effects of water withdrawals on St. Johns River fishes will require, at the very least, coordinating the results of the fish studies with studies on hydrology, benthic macroinvertebrates, and aquatic vegetation. Fish behavior and ecology are influenced both directly by salinity and indirectly by many other factors that also may be influenced by salinity shifts. The cumulative interaction of the assemblage will be critical to understand fish responses to salinity (Peterson, 2003). Furthermore, fishes likely will respond to changes in SAV and macroinvertebrate communities in ways that are entirely independent of salinity changes. The WSIS has yet to integrate the fish work with work on other biological assemblages.

One final issue is the potential change in the landscape habitat mosaic of the St. Johns River (also known as beta- or between-habitat diversity, MacArthur, 1965) that might occur with water withdrawals. Davis et al. (2002) noted that common estuarine fauna were replaced by hard-substrate marine fauna and fishes in areas of mud bottom altered with large granite rip-rap in San Diego Bay, California. In this example, human alteration of the environment changed the estuarine community to resemble an open water marine community. Water withdrawals on the St. Johns River could alter beta-diversity in the basin by causing some environments and their associated communities to disappear (e.g., as in the San Diego Bay example just cited) within segments of the basin. Water withdrawal could also alter the mosaic of environment types in both estuarine and freshwater segments of the basin—changes that could substantially affect resident species and migratory fishes and birds that use mosaic environments to survive throughout the year.

WATER AND NUTRIENT BUDGETS

In reviewing Phase I of the WSIS, the committee often had difficulties placing the proposed water withdrawals into context relative to current conditions in the river. Specifically, the committee found it difficult to obtain answers to such questions as:

(1) What proportion of the total flow at various locations along the river does 262 MGD represent, and
(2) How does this proportion vary in relation to standard hydrologic statistics like annual average, seven-day average low flow at a ten-year recurrence interval (7Q10), and other analogous flows?

The committee is confident that the District has large and detailed historical records of flows at various locations in the river, whereby such calculations can be made, and we encourage the District to do so. Indeed, the District recently informed the committee that the 155 MGD proposed continuous withdrawal at DeLand is approximately 7.8 percent of the average daily flow over the period of record. This is clearly a significant fraction of the average daily flow, and will have increased effects under low flow conditions. Thus, seasonal analysis of the relationship between the proposed withdrawal and the water budget should be provided. On a broader scale, **it also would be useful for the District to assemble basic water budget information for the drainage basin and its major subunits.** These budgets should include the standard sources, sinks, and storage components of such budgets, as well as information on how these components vary over time.

One important use of the water budget information will be to develop nutrient (nitrogen and phosphorus) budgets for key subunits of the drainage basin. As mentioned in more detail in Chapter 1, such information is needed in the WSIS to conduct a credible analysis of the impacts of water withdrawals on algal blooms in the river and especially in the large lakes (e.g., Lake George) that constitute major parts of middle portion of the St. Johns River. Comprehensive information on nitrogen and phosphorus loadings to the system currently is lacking, or at least has not been assembled in a system-wide way. The budgets need to include nitrogen and phosphorus loadings from springs, as it has been noted by District monitoring activities that elevated concentrations of nitrate and phosphate occur in some springs, apparently as a result of human activities in the recharge areas of the Floridan Aquifer. Nutrient budget information also is needed to evaluate the importance of nutrient loadings from the return flows discussed below. The extent to which extended periods of low flow conditions will exacerbate algal bloom conditions in Lake George and other large impoundments of the river cannot be assessed properly without more comprehensive nutrient budget information for the system as a whole.

Quantifying Return Flows

Water withdrawals are treated in the WSIS as consumptive uses with no return flows of withdrawn water to the system. District scientists stated that they used this approach because (1) they do not know where, how, or to what extent withdrawals would return to the system and (2) this is the most conservative case to analyze. The committee appreciates the difficulties involved in estimating future patterns of consumptive water use, given uncertainties regarding future climatic and land-use conditions (e.g., types and extent of agriculture, further development of low-density residential areas, and water-demanding recreational facilities like golf courses). Nonetheless, substantial data are available to the SJRWMD staff on past and current consumptive and non-consumptive uses of water in the drainage basin based on existing patterns of development. **It should be feasible to provide a range of scenarios of the extent to which future withdrawals will be consumptive uses.**

For both hydrologic and water quality budgeting purposes it also is important to predict the extent to which non-consumptive uses will return directly to surface streams versus recharge of groundwater. Return flows of water used for agriculture almost certainly will have elevated concentrations of nutrients, and depending on the agricultural use also may have elevated levels of pesticides, antibiotics (used in animal production), and eroded soils. Return flows from residential uses and such associated uses as golf course irrigation similarly can be expected to be degraded in terms of nutrient concentrations and a wide variety of organic contaminants. Consequently, to the extent that withdrawn water is returned to the river (directly or indirectly), it is likely to return in a degraded condition.

DEWATERING OF FLOODPLAINS AND WETLANDS

Results from the hydrologic modeling studies suggest that the decline in surface water levels produced by the proposed water withdrawal will be small, perhaps 1 cm to at most 4 cm. Initial consideration of this result might suggest that the likely environmental effects will be minimal. The project team should avoid such a judgment as they consider two issues. First, **careful analysis should be completed to determine the floodplain area to be dewatered as a result of the lower water levels, including the timing and duration of dewatering events.** Because of the flat terrain of the St. Johns River Basin as a whole and the extremely flat terrain near the river channel, the river has an extensive floodplain and associated riparian wetlands that are of great ecological significance. The flatness of the terrain also implies that water withdrawals are likely to induce small changes over very large areas. Thus, hydrologic modeling of the floodplain is essential to address questions about water withdrawal impacts on wetland biota and on important biogeochemical cycling processes in floodplain and riparian wetlands. This type of hydrologic modeling was not conducted in the Phase I effort, although it is of critical importance to the biogeochemistry, benthic, fish, and wetlands workgroups. The committee is pleased to find that this modeling will be a part of the Phase II studies (see SJRWMD, 2009c), but it is unclear how this information will be used by the District to address questions about impacts on wetland biota and biogeochemical cycling processes.

Second, an effort should be made to determine the nature and areal extent of locations that may not become completely dewatered but may experience altered biology (e.g., benthic invertebrates, vascular plants, algae and diatoms, soil changes, fish). As discussed in Chapter 4, partial dewatering of wetlands and floodplains may affect species composition and abundance and other ecological characteristics, including production. Direct effects on those areas and from these changes in habitat should be considered. For example, do the larval stages of any fish depend on relatively shallow areas that will be altered by water withdrawals? The Phase I report includes a thoughtful discussion on the differential sensitivity of diverse invertebrate taxa to water level fluctuations. Coupling that discussion with knowledge of the areal extent of habitat changes, especially the floodplain edge microenvironments such as emergent vegetation and woody debris, will be an important component of the next phase of the project.

SUMMARY

In summary, several issues touch on the goals of many of the workgroups and should be considered by each as the WSIS enters its second phase. These include the need for integration of information and results across the workgroups, assessment of future growth in the basin and its relationship to potential withdrawals, creation of water and nutrient budgets for the basin, and more comprehensive investigation of how the floodplains and wetlands in the basin will be dewatered by potential water withdrawals.

3

Hydrodynamic and Hydrologic Modeling

Hydrodynamic and hydrologic modeling of the St. Johns River basin provides a critical foundation for the District's assessment of the impacts of the proposed 262 million gallons per day (MGD) of water withdrawals. Extensive surface water and groundwater modeling was conducted as part of Phase I of the Water Supply Impact Study (WSIS) so that the ecological and biogeochemistry working groups could use the model output to assess potential impacts on biota in the region. This chapter reviews the hydrodynamic and hydrologic modeling efforts and provides recommendations for improvements for Phase II of the study.

SURFACE WATER HYDRODYNAMICS AND HYDROLOGY

In the WSIS Phase I assessment, the surface water hydrodynamic and hydrologic studies focused on understanding the changes in surface water depth, discharge, age, salinity, turbidity and wetland dewatering caused by surface water withdrawals from the St. Johns River. Hydrodynamic and hydrologic changes are the underlying drivers of ecological change resulting from surface water withdrawals; therefore, the hydrologic effects must be predicted accurately to understand the probable ecological effects. An overall project goal is to develop "quantitative response functions" that link river water withdrawal to hydrologic responses and ecological effects. The Phase II Methods report (SJRWMD, 2009c) provides an overview of the hydrodynamic and hydrologic modeling milestones and tasks that will be undertaken as part of the continuing study.

Assessment of Phase I Surface Water Studies

The Phase I surface water hydrodynamic and hydrologic studies (SJRWMD, 2008, Volume 1, Chapters 1 to 4) were screening studies based on readily available science and hydrologic models. Their objectives were twofold: (1) to understand which physical processes are affected by surface water withdrawals and (2) to develop the modeling infrastructure for Phase II studies by building on prior modeling efforts of the St. Johns River Water Management District (SJRWMD or "the District").

The surface water sections of the Phase I report provided an overview of the tools and capabilities that the District had in hand. Historical data were used in Chapter 1 to analyze water surface elevations in the middle St. Johns River (near DeLand). Chapter 2 provided an overview and analysis of hydrodynamic modeling of the lower St. Johns River. Chapter 3 provided information on several salinity scenarios conducted for the lower St. Johns River. Chapter 4 examined a model analysis of possible changes in middle St. Johns River sediment loading. The hydrodynamic models used were well-established three-dimensional models: the Environmental Fluid Dynamics Model (EFDC) in the lower St. Johns River and the Curvilinear Hydrodynamics 3-Dimensional (CH3D) model in the middle St. Johns River.

Three hydrodynamic questions were addressed in Phase I:

1. Under what conditions is the water level at and downstream of DeLand, Florida, dominated by tidal flows rather than river discharge? (Chapter 1)
2. Can the pre-existing lower St. Johns River hydrodynamic model used for total maximum daily load (TMDL) studies be applied to upstream oceanic salinity modeling, and if so, what does it predict for upstream movement of brackish water under different dredging and surface-water withdrawal conditions? (Chapters 2 and 3)
3. Do water level changes in the middle St. Johns River impact sediment resuspension and hence turbidity? (Chapter 4)

The Phase I report adequately addressed these questions for the purposes of a screening study. Specific suggestions on the analyses in support of these three questions follow.

Chapter 1 examined historical hydrological data and behaviors, concluding that water surface elevation (stage) will be relatively unchanged in the lower and middle St. Johns River despite surface water withdrawals. The analysis showed that river stage is not a good indicator of either local hydrologic or ecological effects. The key observation is that the lower and middle reaches of the St. Johns River are tidally influenced under the critical low-flow regimes, such that reverse flows (i.e., upstream flows) will tend to maintain a relatively constant water level regardless of the downstream discharge. A clear stage-height vs. discharge relationship only occurs at higher flow rates when the proposed withdrawals would not be a significant fraction of the overall flow. This fact significantly simplifies hydrological analyses: the critical questions for withdrawals under low-flow conditions in the lower and middle St. Johns River are related to upstream transport of oceanic salinity (in the lower reaches) and changes in baseflow and water age (in the middle reaches). Hydrodynamic modeling of water levels in the middle and lower St. Johns River provided further support for these conclusions (as discussed below). It should be noted that the historical data and analysis methods in Chapter 1 are inadequate to demonstrate conclusively the potential effects of sea level rise on the middle St. Johns River, so future modeling will need to examine how sea level rise may affect upstream propagation of reverse-flow tidal effects. The possibility of climate-change induced sea level rise providing a nonlinear interaction with surface water withdrawals cannot be neglected, and the hydrodynamic models the District has should be adequate for this analysis. The Phase II methods report is not specific in its modeling scenarios, and so it is not clear whether or not sea level rise will be considered.

Chapter 2 demonstrated that the EFDC hydrodynamic model could represent the tidally influenced surface water elevations in the lower St. Johns River. The report compared observed and simulated values, which are provided for the principal tidal component's amplitude and phase (i.e., the M2 tide), hourly water levels, tidal discharge, and salinity. The committee agrees with the report that "the results presented [in Chapter 2] were adequate for the intended use of the model as a screening-level tool for the Phase I study." The District used the tools available to rapidly get robust answers that would help narrow the scope of future studies. The committee also agrees that during Phase II the District should "further refine the model and incorporate additional verification, skill tests, sensitivity tests, and uncertainty analysis." Tasks outlining calibration, verification and uncertainty analyses have been included in the Phase II Methods report; details of the methods used for these have not been provided for review.

Chapter 3 described a limited study of some possible salinity effects associated with changes in sea level, wastewater diversion, surface water withdrawals, and channel dredging in the lower St. Johns River. The District's analysis indicated minor increases (< 1 ppt) of average salt concentration due to withdrawals and higher increases due to channel deepening. This chapter provided large amounts of raw data and graphs but only limited text and analysis. The analysis focused on increases in average salinity and did not discuss extremes. Average salinity changes may not provide a firm foundation for quantifying possible ecological impacts. The need for extreme event data is discussed in the Phase II Methods report for the littoral zone working group (SJRWMD, 2009c, pg. 68, SAV Risk Model paragraph and 69, last paragraph); such data may also be important for understanding other environmental effects. The Phase II Hydrodynamic Methods report included a task for providing simulation results to the other six workgroups, but the details regarding the kind of data and how the groups will interact have not been provided. It will be important for the Phase II modeling and analyses to be tied carefully to the time-space scales of salinity that are important to ecology, and for the hydrodynamic modelers to have a documented process for determining scenarios and data needs for the other six working groups.

Chapter 4 provided information and results from application of the CH3D model to the middle St. Johns River to study sediment transport and resuspension and water level changes (including linkage to the lower St. Johns River for the latter issue). The hydrological analyses from Chapter 1 implied that water level changes would be small due to the low gradient of the middle river and the lack of any coherent stage-discharge relationship at low flow rates. Results from the hydrodynamic modeling in Chapter 4 provided further support for this idea, indicating that water level changes of 1 to 4 cm could be expected for the proposed 155 MGD withdrawal. The maximum modeled change was 4.3 cm at the very most upstream section of the middle St. Johns River. Three scenarios for different distributions of withdrawals at five possible locations were used to examine water level changes using inflow data from 2005. Although this year was chosen for its data availability rather than as representative of an extreme year, the data set includes flow rates below 1800 MGD at all stations for two contiguous months; during this period the proposed withdrawal was 8 percent or more of the instream flow. Similar to Chapter 3, this chapter also included extensive graphs and data but limited analysis. Results presented in Chapter 4 demonstrated that the CH3D model can be calibrated to the middle St. Johns River, and that changes in suspended solids due to withdrawals in the middle St. Johns River should be negligible. Indeed, the findings in Chapter 1 that the water level in the middle St. Johns River is principally determined by sea level rather than discharge supports the conclusions of Chapter 4 that changes in sediment resuspension in the middle St. Johns River should not be significant.

Recommendations for Phase II Surface Water Studies

Based on the committee's analysis of the Phase I report and a preliminary review of the Phase II work plan, the committee has several recommendations for improving the District's Phase II surface water studies. Specifically, the District should (1) work to connect the separate modeling/analysis efforts of Phase I, (2) examine areas that were not studied in enough detail during Phase I, (3) document model calibration and sensitivity, and (4) develop methods to quantify model uncertainty on the time/space scales at which ecological effects may occur. These issues are described in more detail below.

Connecting Hydrodynamic Models and Analyses

The District needs to improve the integration of research questions and physical models across the principal regions of the St. Johns River. In Phase I, there was no linkage between the different physical models, and it was not clear how the results from the physical models would be qualitatively or quantitatively transferred to the ecological models. In particular, the Phase I surface water analyses did not examine how hydrologic changes in specific river sections relate to changes or effects in the rest of the river. Each change within the middle St. Johns River may have cascading influences into the lower St. Johns River and/or upper St. Johns River in hydrodynamics, hydrology, and ecology. For example, a reduction in the net downstream water flux may allow brackish water intrusion further up the lower St. Johns River. Increased upstream extent of reverse flows may change the balance between groundwater intrusion and wetland exchange with the surface water in the upper St. Johns River. An increase in water age may change water quality characteristics (e.g., nutrient concentrations) in the lower St. Johns River. Although the District no doubt understands these issues exist, the linkages need to be identified and studied more explicitly in the Phase II study. The Phase II work plan did not provide any significant detail on connections between hydrology and hydrodynamics.

A critical limitation of the Phase I studies is that the hydrodynamics analyses were not informed by the time–space scales that are important to the ecological processes. Indeed, page 97 of the Phase II work plan indicates that information and data flow is only from the physical modeling to the ecology and does not have a defined feedback path. This one-way information flow may lead to the creation of problems after the hydrodynamic and hydrologic modeling work is essentially complete. As an example, hydrodynamic models can be used to produce detailed data and time-history statistics of salinity inundation; however, modelers often reduce data to simple daily averages and may not store all of the data necessary to later compute more detailed statistics. Thus, to study salinity stresses on submersed aquatic vegetation, the ecological scientists need to define the time-space characteristics of the salinity data that they would like to receive. Physical model development and analyses should be focused on the time-space scales important to ecology rather than the convenient time-space scales for hydrodynamics. This comment should not be construed as a recommendation to implement a monolithic hydrodynamic and ecological model from the estuary mouth to the uppermost reaches of the St. Johns River. However, the choice of model configurations and boundaries should be focused on representing the critical physical and ecological effects of surface water withdrawal rather than simply by the availability of pre-existing models and methods. To some extent, the Phase II work plan shows better integration between the hydrodynamics and hydrology workgroup and the ecology workgroups than the Phase I studies, although there is still limited communication across disciplinary boundaries.

The conceptual framework discussed in Chapter 2 will help improve connections between the models and analyses. By developing a clear hierarchy from the conceptual framework through hypotheses and research questions that link the physics and ecology, modelers can develop a better understanding of the level of modeling necessary to answer the key ecological questions and develop the models accordingly. In particular, better linkages are needed between model extents, grid size, time steps, choice of dimensionality, analysis methods, and the time–space scales over which the physics drive ecological processes.

Issues Requiring Further Study during Phase II

Several important surface water hydrology issues were not studied or explained sufficiently in the Phase I report. The following section identifies specific issues that merit further exploration in the Phase II study.

Modeling Hydrologic Extremes. The District should clarify how well the modeled flow conditions represent expected extremes, particularly drought. The District will need to examine thresholds beneath which water cannot be withdrawn, either due to negative consequences on baseflow and water levels in the upper St. Johns River or upstream propagation of oceanic salt water in the lower reaches. The fact that removal of 155 MGD may not significantly affect water levels in the middle St. Johns River cannot be used to assert that there will be no consequences. The District understands these ideas, but has not always presented them clearly. The St. Johns River system is different than most rivers in that an extended drought with reduced baseflow will not lead to significantly different water levels through the lower and middle reaches, but it will result in upstream propagation of oceanic salt water and/or changes in baseflow and water levels in the upper St Johns River. Thus, consequences of a long-term drought and practicality of water withdrawal in the lower and middle reaches should not be focused solely on water levels in those sections.

Upper St. Johns River Hydrology. The Phase I studies did not include hydrodynamic or hydrologic models of the upper St. Johns River. Thus, in conjunction with improving the existing models, the District should develop a model to represent how surface water withdrawals affect the upstream wetlands and groundwater supply. The District needs to explore whether a pure hydrologic model (e.g., Hydrologic Simulation Program-Fortran [HSPF]) or a combination of a hydrologic model and a one-dimensional river model (e.g., Hydrologic Engineering Center-River Analysis System) or a two-dimensional floodplain inundation model is required to answer the key ecological questions for the upper St. Johns River. This would be determined best by close collaboration with the ecologists regarding the level of detail and certainty needed to understand possible impacts.

Recognizing the above-mentioned deficiencies in the Phase I report, the District proposed in its Phase II work plan to build upon HSPF models previously developed for different purposes for "most of the watersheds." The tasks outlined for this hydrologic modeling effort focus on rationalizing the disparate data sets used for prior models to develop *land use, rainfall,* and *evaporation/transpiration* data sets that are consistent across the models. These tasks appear to be reasonable, although the work plan gives no indication of how the concerns of the environmental science workgroups will be considered in developing scenarios and output data.

Lower St. Johns River Stratification and Salinity Excursions. Vertical gradients of salinity (stratification) and horizontal upstream salinity excursions in the lower St. Johns River need to be more extensively studied during Phase II. There can be little doubt that sufficiently large withdrawals from the middle St. Johns River will allow greater upstream movement of higher salinity water in the lower St. Johns River, which may affect salinity-sensitive habitats. However, the model used in Phase I appears to have too few grid cells in the vertical direction to accurately represent the tendency of heavier saltier water to flow under lighter fresher water. This behavior produces a "salt wedge" that may propagate upstream against the river current based solely on density differences (i.e., a "baroclinic force"). Vertical stratification in the salt wedge reduces mixing of dissolved oxygen downward through the water column and may cause reduced oxygen levels (hypoxia) near the bottom. If a salt wedge develops, the horizontal extent of upstream salinity excursions along the bottom may be significantly greater than indicated by the existing EFDC model or by vertically averaged results.

A technical difficulty in representing salt wedge behavior is that stable numerical models always have some positive numerical diffusion, which is cumulative and tends to smear the vertical salinity gradient over time. The reduced gradient leads to a smaller modeled baroclinic force, which reduces the modeled upstream excursion of salinity. To further complicate the physics, strong wind events may lead to rapid vertical mixing such that the interplay between the stratification, wind-driven turbulence, and current-induced shear requires very careful modeling.

One of the difficulties of hydrodynamic modeling in a salt-wedge environment is validating a model. It is possible to get reasonable values for the average daily salinity while dramatically under-predicting the excursions, maximum salinities, and stratification that may be critical to ecological responses. To quantify how well its model predicts the dynamics of upstream salinity excursions and the development of salt wedge behavior, the District needs data on the evolution of the salinity field over sub-daily time scales. It is suggested that the District undertake several short-term field studies to monitor the spatial and temporal propagation of the salinity gradients in the lower St. Johns River over several tidal cycles. Such studies could be conducted with one or two small boats operating over 48 continuous hours using inexpensive conductivity-temperature-depth, dissolved oxygen profilers. Using these data, the District could examine their model performance with simplified domains to determine the necessary vertical grid resolution required for modeling salinity in the lower St. Johns River.

Effects of Bridges. The lower St. Johns River model did not include the effects of bridges. In particular, the bridge for I-295 has a system of piers that blocks a substantial cross-section of the river. A visual estimate by a member of the committee was that between 10 and 20 percent of the river cross-section may be affected. Because of the shallow river surface gradient through this section, it seems unlikely that the bridge significantly alters water surface levels (at least not within the uncertainty of the model). However, the I-295 bridge (and others) may have significant effects on local velocities, turbulence, and mixing. It might be hypothesized that when the strong salinity gradients are undergoing tidal excursions in the vicinity of the bridge, there is much stronger mixing than when the salinity gradients have passed the bridge or are below it. Thus, there is the potential for significant amplification of the upstream salt wedge excursion if surface water withdrawals cause the salinity front to be pushed further upstream from the bridge (i.e., reducing mixing). Conversely, if the withdrawals lead to the salinity front spending more time around the bridge, then mixing may be increased, ameliorating some of the upstream salinity excursion.

Modeling the Effects of Sea-Level Rise. Sea-level rise will be increasingly important as the century progresses, a fact that the District recognized in the Phase I report. Salinity scenarios corresponding to sea level increases of approximately 6, 10 and 20 cm at Mayport (near the mouth of the estuary) were used as input to the hydrodynamic model. The maximum increase in salinity predicted by the hydrodynamic model was considerably less than 0.5 ppt by 2033 (see Figure 27, Chapter 4, SJRWMD, 2008 draft). The committee believes that this is an excellent approach, but the District should extend this work to incorporate the recent higher-end estimates of sea level rise (e.g., Rahmstorf [2007] predicted rises of 0.4 meters by 2050 and 1.4 m by 2100). This would better quantify the long range impacts on salinity distributions in the lower St. Johns River with regard to future sea-level rise and provide managers with a more complete picture than using only the projections of Titus and Narayanan (1995) and the shorter time horizon of 2033. The worry is that if the 262 MGD are allocated by 2033, managers will have much less freeboard to keep the salt wedge downstream for the rest of the century. A more complete model would help quantify the uncertainty of what upstream influences sea-level rise could pose to future generations.

Documenting Model Calibration, Validation and Sensitivity

The usefulness of any model depends on model calibration and validation. "Calibration" requires adjusting model parameters to obtain a good fit between model results and observed data. "Validation" is the comparison of the model to observed data that were not used in the calibration to see how well the calibrated model predicts an independent data set[1]. For models that are entirely empirical (i.e., based only on observation), the calibration parameter adjustment process is irrelevant as long as the final state is acceptable. However, to evaluate deterministic models that are based on physical principles (such as the hydrodynamic model of the lower St. Johns River), documentation is needed on which parameters are adjusted and by how much. The need for these data can be illustrated by a simple example: some models have a parameter for vertical mixing efficiency that might be adjusted to 1.1 to provide an "improved" model output. Such a setting implies the mixing is "110% efficient" and therefore does more work than the energy available. Models that get the right answer for the wrong reasons are generally less useful (and less believable) than models that are wrong by some quantifiable amount but have physically reasonable parameterizations. Furthermore, models are applied within a framework of incomplete knowledge; for example, forcing conditions are not perfectly known, the model grid resolution is rarely ideal, and the model approximations for turbulence and hydrostatic pressure may affect the model output. The calibration process attempts to account for all of these uncertainties with the recognition that errors in the processes and model representations are interrelated.

To provide transparency and confidence in modeling, the District should carefully document the methods and results of calibration. In particular, the following (at a minimum) should be identified.

1. Data used for calibration,
2. Data used for validation (these data should not be used in the calibration data set),
3. Types of data modified in calibration and the philosophy of their selection,
4. Range of modification for each variable changed,
5. Error measures of some calibrated variables in un-calibrated and calibrated simulations,

---

[1] There is some confusion in the Phase II work plans regarding the use of the word "verification" for the hydrodynamic models. The hydrologic community stands in opposition to the majority of the modeling communities in that the word "verification" is often used in place of "validation." In other modeling disciplines, the word "verification" is applied to the process of testing whether the model is bug-free (i.e., consistent with its desired computational equations) so that a verified model does exactly what it is supposed to do (even if it is wrong!). In contrast, a "validated" model is one that matches the observed data, even if it contains bugs and has not been "verified." The District could make its case clearer by pointing out that the EFDC and HSPF models have been verified previously by developers, and their application to the St. Johns River is validated by comparison to data.

6. Error measures of some non-calibrated variables in un-calibrated and calibrated simulations.

The Phase II work plan includes tasks for model calibration, with some detail on the parameter estimation approach for calibrating the hydrological model. Calibration tasks for the hydrodynamic model are outlined, although there is insufficient information provided to determine whether or not the proposed calibration methods are sufficient.

The governing equations that drive the physics in the model (and in the real world) may have varying sensitivity and robustness to different forms of error. Model sensitivity analyses need to be conducted with a particular focus on the physics that are critical to each study rather than the readily measured values. For example, sensitivity analyses for the vertical grid resolution cannot be based solely on water surface level because it is possible for a calibrated model to create an excellent and robust model of the water surface elevations that is insensitive to the choice of the vertical grid resolution, whereas modeling the upstream motion of salinity may be extremely sensitive to this parameter. The District should demonstrate the sensitivity of each model to various choices in the model setup. The key question is whether or not the model resolution is sufficient for the questions to be answered in light of the variability of the system.

The committee recognizes that it may not be possible for a completely converged model to be applied (i.e., a model that is entirely insensitive to further refinements in time step or grid scale). Such unavoidable sensitivity to model grid scale and time step should be considered within the quantification of uncertainty (discussed below). Sensitivity tests are mentioned in the Phase II work plan as a part of the Task 4 Initial calibration (page 97), but the reference provided is circular (i.e., refers to itself for detail).

Quantification of Uncertainty

Uncertainty in surface water modeling will propagate through associated ecological analyses, adding to the range of possible outcomes and risks. It follows that Phase II should have a strong focus on understanding the underlying sensitivity and uncertainty in the hydrodynamic and hydrologic modeling studies. Two key issues contribute to model uncertainty: the sensitivity of the model setup and unknown future forcing. Both issues need to be estimated quantifiably so that the ecological analyses can be conducted in light of a range of possible future conditions.

In quantifying the uncertainty associated with model sensitivity, the District will need to examine the most recent literature. This area is still relatively new for 3D modeling, and so there may not be clearly applicable guidelines. One aspect the District might consider is the difference between calibrated and uncalibrated model results, which may indicate the maximum range of uncertainty in the model behavior. Another possible avenue is to examine the variance associated with a collection of simulations using different (but reasonable) calibrations.

When quantifying the uncertainty associated with unknown future forcing, it should be recognized that the deterministic hydrodynamic and hydrologic models provide a single answer for a given set of forcing conditions. Since the actual future forcing conditions are entirely unknown, they must be synthesized in some coherent approach. It is possible to use historical data to develop a stochastic understanding of the past for generating possible future "likely" scenarios, but such efforts should be tempered by the realization that we do not have a stationary climate system.

To provide a sound basis for analyzing potential ecological effects, the hydrodynamic and hydrologic models should be used over a variety of scenarios as a means of quantifying uncertainty. In the past, the computational expense of 3D models made such efforts impractical. However, the latest generation of 8-processor workstations allows eight models to be run simultaneously with a relatively modest (<$5K) investment. Thus a bank of five computers can run 40 simulations at a time and rapidly develop a data set large enough to quantify the uncertainty associated with the different possible future conditions. To make this analysis tractable, the District should develop a rigorous approach that generates a variety of possible future scenarios and evaluates their likelihood, along with methods to analyze results from multiple model scenarios in an uncertainty framework. The Phase II work plan includes an Uncertainty Analysis Plan, apparently to be conducted by INTERA Geosciences & Engineering, but no details were provided.

Summary of Surface Water Hydrodynamics and Hydrology

The District is progressing along the correct track, but critical details discussed above either have not been considered or have not been sufficiently documented. The committee has three specific recommendations. First, the District should connect the separate modeling and analysis efforts of Phase I. For example, the Phase I analyses did not examine how hydrologic changes in specific river sections relate to changes or effects in the rest of the river. Second, several areas were not studied or explained in enough detail during Phase I, including modeling of extreme conditions, the hydrology of the upper St. John River, vertical gradients of salinity and horizontal upstream salinity excursions in the lower river, and the effects of bridges and sea level rise. Third, the District should document model calibration and sensitivity and develop methods to quantify model uncertainty on the time/space scales at which ecological effects occur.

GROUNDWATER HYDROLOGY

A primary goal of the Phase I groundwater modeling was to predict whether discharges of groundwater into the St. Johns River would change if water withdrawals caused the river stage drop. Two groundwater flow models were used during Phase I to compute groundwater base flows along the river from the surficial aquifer system (SAS) and the upper Floridan aquifer (UFA): the North Central Florida (NCF; Motz, and Dogan, 2004) and the East Central Florida (ECF; McGurk and Presley, 2002) MODFLOW (Harbaugh et al 2000) models (see Figure 3-1; Table 3-1). These steady-state models are not density-dependent groundwater flow models, such that solute transport is not addressed. The groundwater analyses were confined to the middle and upper St. Johns River basins (see Figure 1-1); for the purposes of this analysis, the District assumed that there is no groundwater discharge contribution to the lower St. Johns River basin.

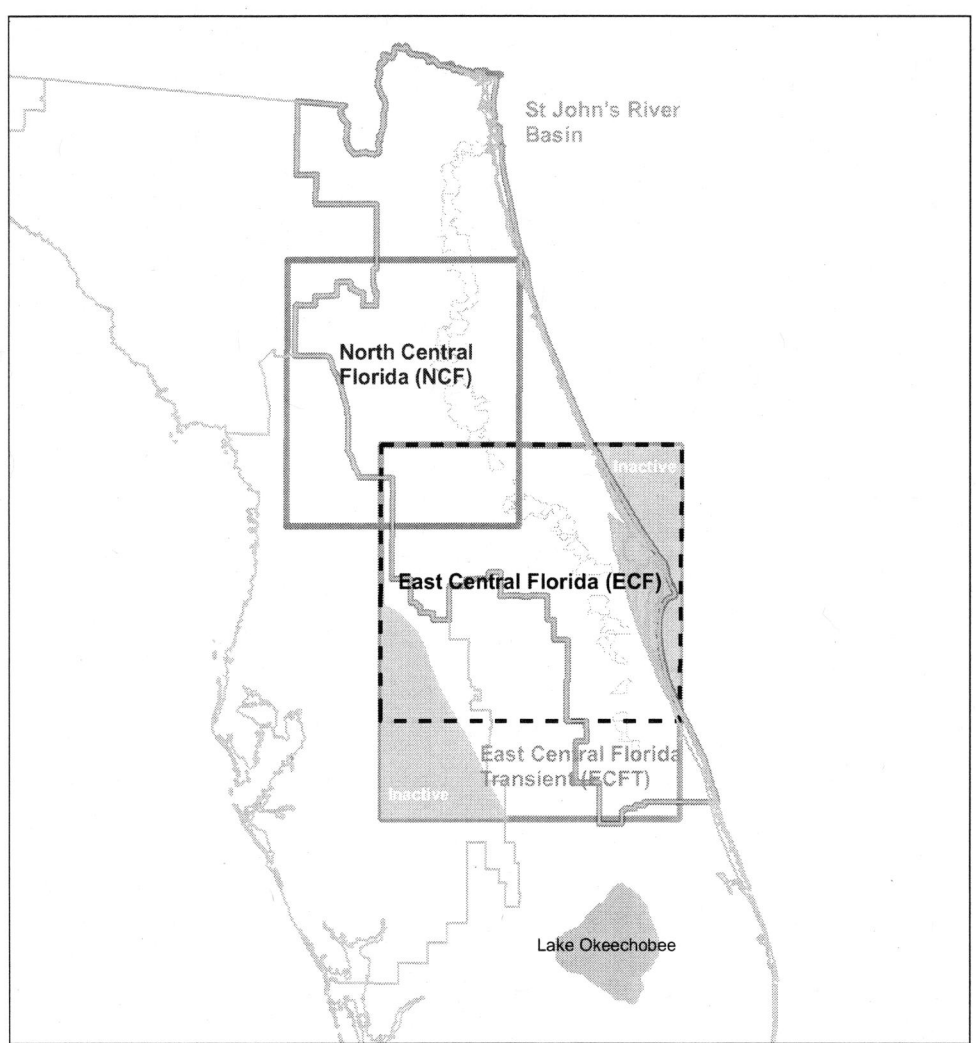

FIGURE 3-1: SJRWMD Groundwater Models used in the WSIS.
SOURCE: SJRWMD.

TABLE 3-1: Groundwater Models in the WSIS Area

Model Name	Model Area (Status)	Row, Column	Uniform Square Grid Size	Flow Type	Aquifers	Authors
East Central Florida (ECF)	Mainly the upper St. Johns River Basin (Used in Phase I CIA)	174, 194	2,500 ft on a side.	Steady State	SAS, UFA, & LFA	McGurk & Presley (2002)
North Central Ground Water Model (NCF)	Mainly the middle St. Johns River Basin (Used in Phase I CIA)	150, 168	2,500 ft on a side.	Steady State	SAS, UFA, & LFA	Motz et al. (2004)
East Central Florida Transient Model (ECFT)	Whole upper St. Johns River and Kissimmee Basins (Not Used in Phase I CIA)	472, 388	1,250 ft on a side	Transient	SAS, UFA, & LFA	Butler et al. (2009)

Note: SAS = surficial aquifer system, UFA = upper Floridan aquifer, and LFA = lower Floridan aquifer.

Assessment of the Phase I Efforts

Based on the modeling results, the change in average discharge of groundwater following water withdrawal was not predicted to be significant in the middle and upper basins. More specifically, a 4-cm decrease in water level would increase the groundwater base flux by 1.1 percent (or a similarly small volume); it would increase the chloride flux by 13 million tons/day or 1.2 percent (SJRWMD, 2008). To further demonstrate that the groundwater discharge variability is not significant along the river, the District conducted a modeling analysis of two extreme cases. In the first case, the model results showed that lowering the water stage drastically (by 30.5 cm) would nearly double the groundwater discharge (which implies that a lowering of 4 cm stage would only change groundwater discharge by a small percentage, assuming a linear relationship). In the second case, the model predicted that if the current groundwater pumping in the St. Johns River basin (about 700 MGD) was stopped, the groundwater base flow to the river would increase to about 30 MGD, which is 5 percent of what is pumped (suggesting that discharge is not particularly sensitive to groundwater withdrawals). (The model results in the second case did not consider the direct surface water runoff that would increase as water moves from wetlands and other areas toward the river.) From these analyses, the District appears to have concluded that groundwater discharges to the river are relatively insensitive to the current proposed changes in river stage. This conclusion may be correct, but it is not yet technically defensible, for the reasons discussed below.

Limitations of the Current Groundwater Modeling

Several limitations in the current groundwater models may affect the accuracy of the output. First, the models are steady state models, they do not model density-dependent groundwater flow, and wetland hydraulics is not represented in any of the groundwater models. These facts are not surprising, given that it would be extremely difficult to develop a rigorous regional model within the timetable of the WSIS.

To elaborate, the Floridan aquifer system of northeast Florida, mainly the middle St. Johns River basin (see Figure 3-1), was simulated using a steady state regional numerical model (the NCF). This revised groundwater flow model (Motz and Dogan, 2004) includes a change with respect to previous versions of the model in its lateral boundary conditions and recharge and evapotranspiration calculations. The newest design of the NCF includes all groundwater withdrawals for the 1995 conditions in the surficial aquifer system and upper Floridan aquifer. Vertically, the model consists of three aquifer units, separated by confining units. For both the NCF and the ECF, model calibration was based mainly on piezometric heads and spring flows. However, a rough calibration of the base flow using the average mass-balance also was conducted on a larger scale for the steady-state models. The models are not capable of computing changes in groundwater base flow resulting from the variability in rainfall conditions, nor are they capable of predicting saltwater intrusion into the river because variable-density flow cannot be accounted for.

The ECF model (McGurk and Presley, 2002) was developed mainly for the upper St. Johns River basin to simulate the freshwater portions of the surficial and Floridan aquifer systems and to assess the potential impacts of future withdrawals to the Floridan aquifer (see Figure 3-1). The District calibrated the ECF model to average 1995 conditions by comparing the simulated groundwater levels to the surficial aquifer and the Floridan aquifer systems' historical groundwater levels, and by comparing simulated spring flow conditions to the actual ones from 1995. McGurk and Presley (2002) recommended the following to further improve model results: calibrate the model to transient conditions, gather and use more information for recharge and evapotranspiration into the surficial aquifer system, and evaluate saline water intrusion. The District has not acted on these recommendations yet, which would help create a more defensible model and bolster their conclusions about the potential effect of water withdrawal on groundwater levels.

Finally, the District's 20-year groundwater projections do not include climate change factors, which would be a difficult and challenging difficult task. Sea level change is addressed in the Phase I report only in terms of the projected increase in salinity of the lower river.

Adequacy of Groundwater Inputs into Surface Water Models

The above limitations have direct consequences for the District's hydrodynamic modeling because the groundwater model results (i.e., groundwater base flow from the aquifer) are used as input into the hydrodynamic models. Because the groundwater models are steady state, the salinity sources into the river are considered to be constant over time. However, rising seawater levels, intense droughts, and/or increasing withdrawals from the river could eventually change the contribution of groundwater to the river's overall salinity. Data show that there are significant amounts of chloride in the groundwater that may make their way into the river (see Figure 3-2). The salinity in the river, specifically in Segment-7 (Figure 3-3), is dependent on geologically old saline water (around 5000 ppm) stored in the upper Floridan aquifer that has not been completely flushed out by meteoric water. (Indeed, the last opportunity for the aquifer system to have drained the saline water was approximately 18,000 years ago when the sea level was dramatically lower [by 450 ft.] than today.) The degree of uncertainty about salinity coming from this source was not evaluated in Phase I of the WSIS.

Recommendations for Phase II Groundwater Studies

Based on the committee's analysis of the Phase I report and a preliminary review of the Phase II work plan, the committee has several recommendations for improving the District's Phase II groundwater study. Improvements to the groundwater models during Phase II of the WSIS should include incorporating transient groundwater flow, two-dimensional density dependent flow, and the interactions between the wetlands and the aquifer system. These issues are described in more detail below.

Transient Groundwater Flow

A regional-scale, transient groundwater model covering east-central Florida, called the East-Central Florida Transient (ECFT) model, is in the final stages of development (Butler et al., 2009). ECFT expands upon the ECF model to include the Kissimmee Basin (see Figure 3-4).

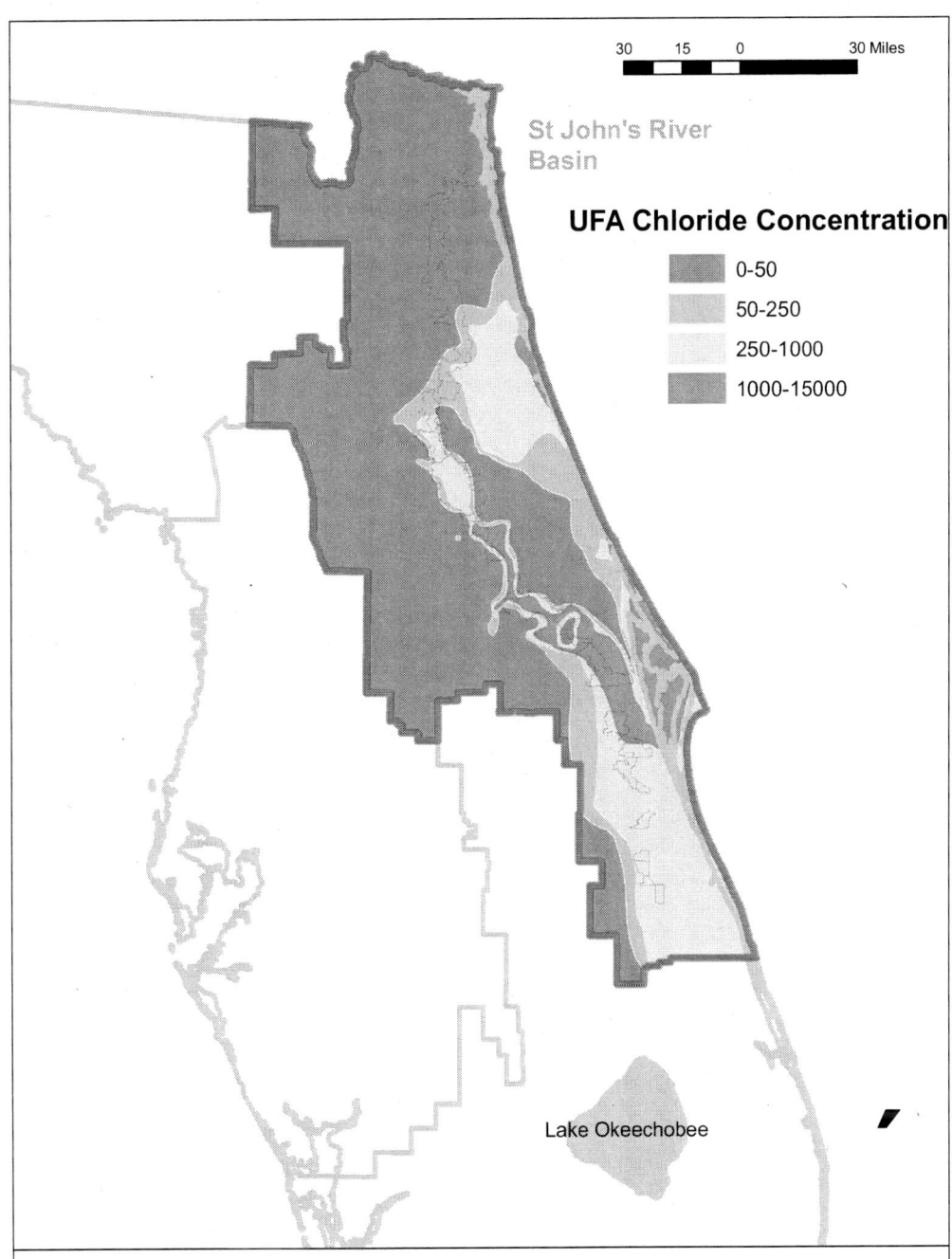

FIGURE 3-2: Generalized Salinity Map, Layer 2, ECFT Model.
SOURCE: SJRWMD Staff.

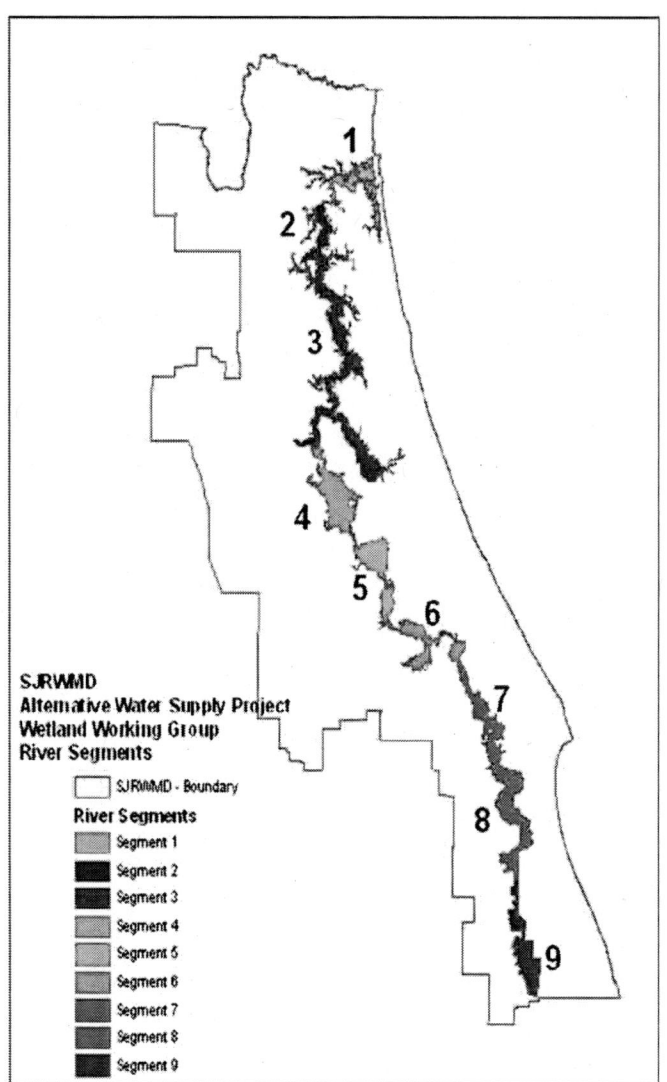

FIGURE 3-3: St. Johns River Segments.
SOURCE: SJRWD.

The St. Johns River, Southwest Florida, and South Florida Water Management Districts are all in the final phase of calibrating the model under transient conditions. Following calibration, verification, and incorporation of peer reviewed comments, the District plans to use the model to evaluate future impacts on the hydrologic system, including wetlands, lakes, and springs. Use of this model will be an important step in the right direction for the Phase II WSIS.

ECFT includes a Wetlands package, developed by the South Florida Water Management District and the Center for Hydrology and Water Resources at Florida Atlantic University (Restrepo et al., 1998). The current version of the Wetlands package, which has been applied to subregional models in south Florida, is incorporated into MODFLOW and enables the top layer of the grid system to contain overland or groundwater flow. Such an approach allows the flow equations to remain valid when the water surface falls below the soil surface. Furthermore, the package can account for vegetation characteristics, sheet flow, sloughs, levees, barriers to overland and groundwater flow, and evapotranspiration. Thus, it is suitable for modeling the wetting and drying of wetlands.

The ECFT grid spacing (see Table 3-1) balances the need for resolution of surface water features and local impacts with data availability. The temporal discretization of the model (the process of transferring continuous models and equations into discrete counterparts over a specific time step) was chosen to accurately reflect the hydrologic system changes (e.g., rainfall and canal stages) over a recent wet and dry cycle. This model should be better able to predict wetlands behavior and interactions with the river, the aquifer, and the surface water storage. It will be critical to simulate the wetlands behavior in the specific locations where the water withdrawals may be occurring. ECFT should be a valuable screening tool because of its transient qualities, including the representation of various seasonal changes, wetland simulation capabilities, and increased horizontal resolution.

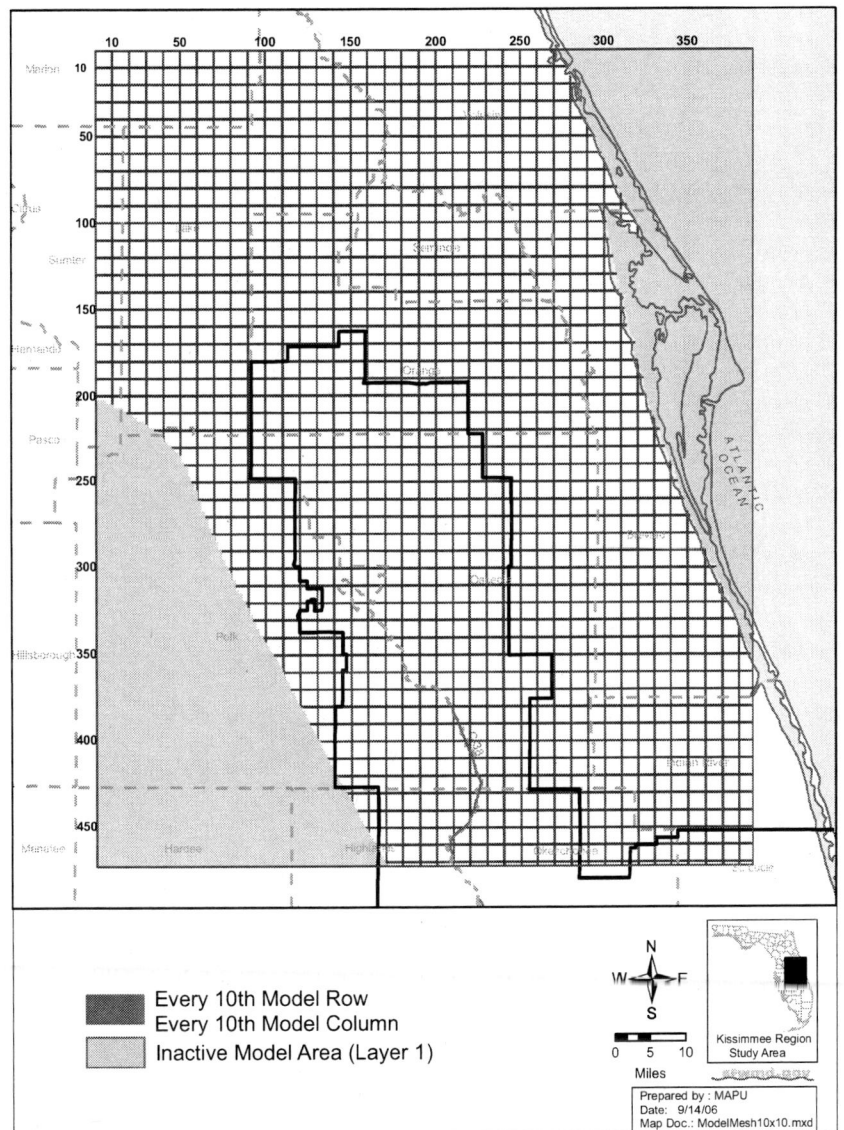

FIGURE 3-4 Model Study Area and Grid for ECFT.
SOURCE: SJRWMD.

Density-Dependent Groundwater Flow

The ECFT is not a density-dependent model, which limits its ability to predict changes in salinity that might result from water withdrawals. A density-dependent groundwater flow and transport model remains an important need for the St. Johns River basin. It would be extremely challenging to develop a three-dimensional model within the timeframe provided for the WSIS, but as an alternative the District should consider developing a groundwater flow and transport model for a cross-section of the basin, preferably in river segment 7 (see Figure 3-3) because that is the most critical location for potential changes in salt water intrusion. Seepage from the aquifer to the river in segment 7 is significant because the confining unit between the bottom of the river and the upper Floridan is thin in that zone, and the head in the upper Floridan is higher than the river stage. The gradient in the upper Floridan decreases significantly toward the ocean.

Such a cross-sectional model could indicate the vulnerability of river segment 7 to changes in saline flow from groundwater following potential increases in water withdrawals from the river or other alternative water management scenarios. Such an effort is worthwhile and feasible because the problem of saline intrusion is density-dependent, the ideal tools to analyze the problem do not yet exist, and a two-dimensional cross-sectional model should not take long to develop. Indeed, there is no groundwater pumping from the upper Floridan aquifer in river segment 7 (because of high salinity), which simplifies the development of a cross-sectional model. Cross-sectional models are well suited for performing sensitivity analysis and can simulate groundwater flow patterns at a refined level of spatial resolution near the river locations of interest.

Modeling Wetland-Aquifer-River Interactions

As mentioned in Chapter 2, when water is taken out of the river, several hydrologic results can occur, including reduced flow downstream, increased base flow to the river due to a steeper gradient between the aquifer and the river, and increased drainage from upstream wetland areas. To analyze these occurrences, it is necessary to use an integrated surface water–groundwater flow model that is able to represent wetland and aquifer interactions. The current models (NCF and ECF) are not able to quantify the flow within wetlands and their change in storage. The Wetland package of ECFT is able to simulate these interactions sufficiently, although the committee has not yet seen the details regarding calibration of the ECFT model.

Summary of Groundwater Hydrology

In order to improve the groundwater modeling in Phase II of the WSIS, the District should consider using the ECFT transient model, which includes wetlands processes, and a cross-sectional density-dependent model. These models will be critical to understanding how groundwater flow and salinity flux into the river will vary with water withdrawals. The more sophisticated analyses that would derive from these models are needed to support the District's assertion that water withdrawals will not significantly alter the groundwater discharge contribution to the St. Johns River basin.

4

Ecological Assessments

In the Phase I study, six St. Johns River Water management District (SJRWMD) workgroups used hydrodynamic and hydrologic modeling data, existing monitoring data, and literature reviews to provide preliminary assessments of potential biogeochemical and ecological impacts in the St. Johns River from withdrawing 262 million gallons per day (MGD) of surface water. These efforts will be expanded in Phase II to include additional modeling and data collection to delineate environmental effects boundaries and assess the potential for crossing response thresholds at various levels of continuous withdrawal. This chapter provides assessments of the Phase I report chapters on biogeochemistry; plankton, nutrients, and total maximum daily loads (TMDLs); the littoral zone; benthos; fish; and wetlands and wetland-dependent species. For each workgroup, the chapter includes recommendations to improve the Phase II efforts.

BIOGEOCHEMISTRY

The biogeochemistry workgroup of the Water Supply Impact Study (WSIS) identified seven potential effects of additional water withdrawals on biogeochemical processes and conditions in the St. Johns River drainage basin. All seven effects are related to the possibility that soil accretion will be reduced and/or oxidation of organic soils (histosols) will be enhanced in riparian wetlands associated with the extensive floodplains of the St. Johns River as a consequence of changes in river stage induced by additional water withdrawals. The floodplains have swampy herbaceous wetlands with deep organic soils, and withdrawals could increase the number of days that organic wetland soils are exposed to air. Exposure promotes oxidation and diagenesis of organic matter and release of various substances that can be exported to the river when the soils are inundated again.

During Phase I, the workgroup concluded that the potential effects of changes in pH and increased releases of organic inhibitors, labile dissolved organic carbon, and metals from oxidizing organic soils had unknown significance. However, three other potential effects were considered to have potentially high significance: (1) reduced nutrient sequestration, (2) increased release of colored dissolved organic matter (CDOM), and (3) increased production and reduced sequestration of greenhouse gases (carbon dioxide, methane, nitrous oxide) produced within inundated organic soils. The workgroup decided to focus on nutrient and CDOM release because they were thought to have the greatest potential for effects on the river.

Assessment of Phase I

The Phase I study involved calculations based on literature values for release rates of various constituents from flooded and exposed (i.e., dewatered) organic soils. No new field-based measurements or studies were performed. The additional duration of soil exposure was estimated for nine segments of the middle and upper St. Johns River basins based on hydrologic stage-frequency curves, and the areas of organic soils that would be affected were estimated from geographic information systems (GIS) data. Riparian wetlands in the lower St. Johns River were not included in these calculations because stage is controlled in that region by tide rather than river flow. The incremental loading to the river was based on a mass calculation of the change in release rates of substances between the inundated and drained states:

$$\Delta L = (FD - FI) \times A \times T$$

where ΔL is the change in load to the river (g yr^{-1}), FD is the areal release rate when drained (g m^{-2} d^{-1}), FI is the areal release rate when inundated (g m^{-2} d^{-1}), A is the affected area of organic soils (m^2), and T equals the additional days of exposure (d yr^{-1}). A positive change in loading indicates increased release of nutrients or organic matter to the river (or atmosphere), and a negative change indicates increased retention. Implicit in the above equation is the assumption that net primary production in the plant community associated with affected organic soils does not change between the inundated and drained states. The approach thus assumes that net change in flux is a reflection solely of changes in oxidation rates between inundated and drained states. The committee is concerned that no supporting evidence is provided for this assumption.

Preliminary results were presented in the Phase I report for inorganic phosphorus release only. The authors suggested that increased releases caused by additional exposure (dewatering) of organic soils could be significant for Lake Winder (in the upper St. Johns River) relative to a current TMDL that requires a substantial *decrease* in phosphorus loading from the lake. No results were presented for the other major nutrient (nitrogen), or for CDOM, labile and inhibitory dissolved organic carbon (DOC), and metals because the workgroup concluded that insufficient information was available in the literature on release rates of these substances from exposed histosols. However, the committee notes that Table 1.1 of the Phase I report appears to include as much information on nitrogen release rates from flooded and drained peat lands as it does for phosphorus. The committee concludes that the focus on phosphorus in the Phase I studies was as much a result of the short amount of time available to the workgroup as a reflection on the adequacy of the release rate constants for various constituents. The work plan for the Phase II study indicates that these other substances will be addressed by undertaking appropriate data collection, and the committee concurs with this decision.

The committee discussed at considerable length the issue of potential additional CDOM loadings as a result of increased water withdrawals, and it concluded that the Phase I report did not provide sufficient evidence that this would occur. The committee accepts the *possibility* that under some circumstances increased water withdrawals *could* lead to substantial increases in drying and oxidation of wetland soils and this could lead to losses in organic soils and increases in release of CDOM. Nonetheless, a convincing case was not made that this *would* happen in the St. Johns River as a result of the proposed additional water withdrawals. Oxidation of air-exposed organic soils degrades high molecular weight (insoluble) humic material into smaller, soluble humic and fulvic acid molecules that are exported to streams and rivers during soil reflooding, and chemical transformation of CDOM continues to occur under the influence of ultraviolet light and microbial processes in aquatic environments. Such processes already occur to a substantial extent in the St. Johns River basin. The river and its tributaries and associated lakes are highly colored with humic substances (CDOM) under existing conditions. The Phase I report did not make a persuasive case that changes in CDOM concentrations or loadings under conditions of additional water withdrawals would have significant ecological or water quality impacts. If water flow rates decrease and CDOM loadings remain constant (or increase), the concentration of CDOM would increase, but the Phase I report did not provide evidence that increased water withdrawals would enhance generation of CDOM in the riparian wetlands or that flows during reflooding events would be lower than under present conditions. Nonetheless, the committee recognizes that increased CDOM concentrations could affect light penetration (even though it already is low in this highly colored river) at least for short periods of time. Because primary production in the river may be light-limited (this is especially likely for submersed aquatic macrophytes), this potential effect deserves further attention.

None of the literature values of substance release rates from drained and inundated soils used in the Phase I study (Table 1.1 of the biogeochemistry chapter) appears to be from soils in the St. Johns River basin. Many of the rate constants are from studies in South Florida, specifically the Everglades Agricultural Area (EAA), an area heavily affected by agricultural activities for many decades. Nutrient release rates from EAA soils are not likely to be applicable to riparian organic soils in the St. Johns River basin, which have not been heavily impacted by agriculture and other human activities. There also are many questions regarding the environmental conditions under which release rates reported in the literature were measured and whether those conditions are representative of conditions that would occur if St. Johns River basin organic soils were subjected to additional drainage.

Recommendations for Phase II

The workgroup intends to address the shortcomings described above regarding substance release rates from drying soils in the Phase II studies. To the extent that experimental studies are undertaken to obtain nutrient and CDOM release rates from drained soils, the committee encourages the workgroup to use procedures that will yield data reflective of environmental conditions.

The committee understands that the Phase II biogeochemistry studies will address the question of enhanced nitrogen release from drained organic soils, and it agrees that they should, in addition to enhanced phosphorus release. It is essential that these release rates be placed in the proper context that can come only from having knowledge about the total nitrogen and phosphorus loadings to the river—specifically to the major lakes along the river channel. This information also needs to be obtained and analyzed at a sufficiently fine temporal scale to be relevant in assessing impacts on algal blooms; in a highly dynamic system like the St. Johns River, algal blooms are not controlled by total annual nitrogen and phosphorus loading, but by loadings at critical times of year, particularly those during and preceding low-flow conditions during warm weather periods. Furthermore, the ratio of concentrations of these two limiting nutrients is an important determinant of phytoplankton community composition (Lowery, 1998; Conley, 2000).

The committee appreciates the difficulties and complexities involved in quantifying the additional loading of nutrients, metals, and natural organic substances to the St. Johns River that could result from the sequence:

Additional water withdrawals → Changes in water stage → Increased exposure of soil organic matter to air → Increased oxidation and release of substances → Increased transport to the river upon reflooding.

In addition to large uncertainties in translating predicted changes in stage-duration to area-duration changes in exposure of soil organic matter to air, there are important uncertainties in the extent to which substances produced during air exposure will be transported to the river. In addition, there are substantial issues with regard to predicting the actual extent of substance oxidation and release during exposure to air.

The District plans to conduct laboratory experiments during Phase II to determine substance release rates from drained soils (SJRWMD, 2009c, pp. 7-9). However, the committee is concerned that small-scale (laboratory) experiments will not be able to simulate field conditions adequately. For example, organic soils in field conditions probably receive moisture from upward capillary flow on a continuing basis even when there is no standing water above the soil surface. Mimicking this phenomenon in the laboratory could be difficult. Temperature, duration of drained conditions, and the degree of drying are likely to vary temporally under field conditions. These variations will affect oxidation rates in ways that would be difficult to duplicate in the laboratory. In this context, in situ mesocosm studies indicated in the work plan (p. 8) take on special importance. The work plan does not provide information on the size of the mesocosms, but Figure 2 (p. 9), which is a sketch of the proposed mesocosms, suggests they will be quite small.

Given the well-known heterogeneity of soils, it is probable that many samples will need to be tested. The Phase II work plan provides various numbers in this regard: 150 general sites will be sampled; 36 soil cores will be collected from 12 sites; *in situ* studies will be conducted at an unspecified subset of the 12 sites. Unfortunately, no rationale is presented in the work plan for these numbers.

Finally, there are likely to be important "scale factors" complicating the interpretation of results from laboratory microcosm experiments. For example, the presence of vegetation and extensive root systems in field soils may change soil moisture conditions, as well as the extent of oxic conditions within surficial soil layers. The committee encourages the District to carefully consider these issues in designing experiments so that the results can be of more than academic interest and in fact can be useful in predicting changes at the spatial and temporal scales of interest in the impact assessment.

The above concerns lead to the following recommendations regarding Phase II biogeochemical studies. First, it is critically important that sufficient samples be analyzed to address the well-known large heterogeneity found in soils. If the proposed sampling design was not based on the known or expected diversity of soil types and within-plot heterogeneity of the study area, District staff should revisit their sampling plans. Second, experimental studies should be done at as large a spatial scale as possible to avoid artifacts caused by trying to extrapolate results from small sample sizes and small containers to the ambient environment. Specifically, samples incubated to assess effects of drying on rates of soil oxidation and nutrient release must include roots and vegetation and not just peat soil. In this regard, the committee is not convinced that samples of the size implied in the work plan for Phase II are adequate to give realistic and representative results for soil oxidation and nutrient release from riparian wetland soils in the St. Johns River under the conditions likely to be engendered by enhanced consumptive-use water withdrawals.

Finally, the committee is not convinced that the Phase I studies have provided a sufficient predicate for the laboratory and *in situ* studies proposed for Phase II. In particular, it is premature to conduct extensive laboratory and field experiments to evaluate rates of nutrient and CDOM release from drying soils of riparian wetlands in the St. Johns River Basin because the Phase I studies were not adequate to demonstrate that drying and oxidation will occur on a spatial scale to cause significant impacts to the river ecosystem with the proposed additional water withdrawals. The committee is *not* concluding that such impacts will not occur, nor is it implying that additional water withdrawals will not result in drying and oxidation of riparian soils. Rather, the committee concludes that the Phase I studies were inadequate to demonstrate that this represents a likely impact of sufficient magnitude to warrant further study and analysis. Consequently, the committee concludes that a sequential approach would be more effective and efficient for the Phase II studies. First, additional analyses and experiments need to be undertaken to establish the areal extent of wetland soils that would be dried *to a sufficient extent* and *for a sufficient duration* to enhance oxidation of soil organic matter and subsequent release of nutrients and CDOM. The duration of drying needed to enhance oxidation likely will vary with season, and so timing of drying events needs to be considered. Second, if these studies indicate that further studies are warranted on transformation rates of soil organic matter during drying, then it would be appropriate to undertake experimental studies at mesocosm rather than microcosm scales to measure rates of nutrient and CDOM release under environmental conditions relevant to the proposed water withdrawal scenarios. Studies at this larger (mesocosm) scale are recommended to minimize experimental artifacts and enhance the ability to extrapolate results to the actual St. Johns River system.

Summary

Although the biogeochemistry workgroup identified and ranked several potential effects of additional water withdrawals on biogeochemical processes, their analysis did not use local release rates of constituents from flooded and exposed organic soils. Results were only presented for inorganic phosphorus release, although the work plan for Phase II indicates that nitrogen will be addressed. If laboratory experiments are undertaken to obtain local nutrient and CDOM release rates from drained soils, the workgroup should use sufficient sample numbers to address the well-known large heterogeneity found in soils. Experimental studies should be done at as large a spatial scale as possible to avoid artifacts caused by trying to extrapolate results from small sample sizes and small containers to the ambient environment. Finally, the District should consider the worth of collecting such data before there is better information on the areal extent of wetland soils that would be substantially dried due to water withdrawal.

PLANKTON, NUTRIENTS, AND TMDLs

The plankton and TMDLs workgroup was tasked with identifying and quantifying possible environmental impacts of water withdrawals on plankton communities and existing TMDLs in the lower and middle St. Johns River. In the Phase I report (SJRWMD, 2008), the group attempted to address four principal questions related to plankton and TMDLs using *existing data*:

1. What are the potential impacts and which impacts are likely significant?
2. What mechanisms or empirical relationships connect direct hydrological consequences of

water withdrawal to direct or indirect environmental impacts?
3. What environmental and hydrologic criteria are appropriate to measure impacts?
4. What environmental and hydrological boundaries define significant adverse impacts?

In Phase II, the workgroup will complete the evaluation of potential effects on plankton and related conditions in the river and evaluate specific water withdrawal scenarios. According to SJRWMD (2009c), three main tasks will be to: (1) characterize relationships between hydrologic variables and direct and indirect impacts on plankton, (2) establish environmental boundaries for levels of adverse effects, and (3) identify hydrologic regimes compatible with environmental boundaries. The primary objective is to determine acceptable limits of change (environmental boundaries) in plankton and nutrient-related conditions that are affected by water withdrawals.

Assessment of Phase I

The strategy for addressing the potential impacts to plankton and TMDLs from reduced discharge is logical, sound, and clearly presented, and Table 4 in SJRWMD (2008) provides a comprehensive overview of the progress made in Phase I of the WSIS. Overall, the workgroup did a commendable job summarizing and interpreting archival data from a variety of studies conducted in the middle and lower reaches of the St. Johns River over the past 25 years. However, during Phase I the District was unable to satisfactorily answer the four main questions listed above because several critical issues were not considered.

The Phase II studies need to be designed to address the following critical issues. First, additional water withdrawals may increase the likelihood, duration, and areal extent of water column density stratification and induce bottom water hypoxia in the lower St. Johns River under low flow conditions, which was not discussed in the Phase I report. Such conditions, even if they occur for only short periods of time, could have major effects on water quality in the river, in addition to affecting benthic organisms, nekton, and nutrient cycling processes.

Second, the Phase I report does not adequately address the type or frequency of additional water quality and biological monitoring data needed to adequately assess the impacts of water withdrawals on TMDLs and plankton. This suggests the need for a comprehensive monitoring strategy and protocol to assess the potential impacts of water withdrawal to support the Phase II efforts. Such a strategy would include a detailed list of critical variables to be measured, as well as the frequency and locations of measurements. According to SJRWMD (2009c), Phase II plans neglect large segments of the river that could be affected by water withdrawals by including monitoring only in Lake George. More comprehensive monitoring during Phase II would include key freshwater and estuarine regions and encompass benthic meroplankton (mentioned later in conjunction with the benthos workgroup as an important group of taxa to investigate).

Third, based on presentations by District staff at the January 2009 meeting and on various statements in the Phase I report, it appears that the District has assumed that phosphorus is the limiting nutrient for algal growth in the freshwater portions of the St. Johns River. That assumption may not be valid. The occurrence of substantial blooms of nitrogen-fixing cyanobacteria in Lake George, for example, is strong evidence that nitrogen limitation occurs at least at some times of year and in some locations within the river.

Ecological Assessments

Recommendations for Phase II

One important issue of concern that cuts across the nutrient and plankton subject areas is whether additional water withdrawals would increase the likelihood of water column density stratification and bottom water hypoxia in the lower St. Johns River under low flow conditions. Analysis of this issue in Phase II will require close collaboration with, and careful consideration of the findings from, several other workgroups including the hydrodynamic and hydrologic and biogeochemistry workgroups. Because of the short timeframe of the Phase I study and the fact that the groups were working in parallel to produce their Phase I documents, integration of findings across the workgroups may have been fragmentary and incomplete. Strong interactions between the workgroups must be established in Phase II so that the findings related to the effects of water withdrawals on the physical and chemical driving forces can be thoroughly integrated into the analysis of plankton dynamics. As discussed in greater detail below, more attention should be given to the role of CDOM, heterotrophic bacterioplankton, and microbial loop processes in the analysis of plankton dynamics in the Phase II study.

The effects of high concentrations of CDOM on phytoplankton ecophysiology were not explored during Phase I. CDOM alters the light environment and reduces the amount of useful light available for photosynthesis. Although Secchi depth was included in the Phase I analysis, it is at best a crude indicator of light attenuation (Kirk, 1994). Under conditions of increased water withdrawals, CDOM concentrations could increase if export of CDOM from riparian wetlands remains constant (or possibly increases) and river flow is decreased. This could lower light levels, possibly limiting phytoplankton and benthic macrophyte production. For low-light acclimated phototrophs, small alterations in ambient irradiance can result in exponential changes in rates of photosynthesis. The combination of change in both light quality and quantity due to changes in CDOM concentrations may have a variety of effects on plankton and associated conditions in the river and its lakes.

The potential effects of water withdrawal on heterotrophic bacterioplankton and microbial loop processes also were not addressed in the Phase I report. Given the high CDOM and nutrient concentrations in this system, bacterioplankton likely play an important role in oxygen dynamics, water quality, and biogeochemical cycling in the river (Tranvik, 1990; Schultz, 2000; Joint et al., 2002; Alonso-Saez et al., 2008). Although data on this community in the St. Johns River may not be available, the potential implications of water withdrawal should be explored using data from similar river systems.

Nutrients

Historical data for this system clearly demonstrate that the ratios and concentrations of nitrogen and phosphorus are important determinants of phytoplankton biomass and community composition. System biological responses to water withdrawal will depend directly on the concentrations and ratios of these primary limiting nutrients. In Phase II, there needs to be an explicit differentiation between discharge, nutrient concentration, and nutrient loading. For example, as flow decreases, nutrient concentrations should increase due to lower dilution rates. Will nutrient loadings remain constant while nutrient concentrations increase? As mentioned in the section on biogeochemistry, this information needs to be obtained and analyzed at a sufficiently fine temporal scale to be relevant in assessing impacts on algal blooms, which are controlled by nitrogen and phosphorus loading at certain times such as low-flow conditions during warm weather periods. Thus, the Phase II efforts should emphasize the importance of both nitrogen and phosphorus and the effects of water withdrawals in the context of a dual nutrient management strategy (Paerl, 2009). The Phase II work plans do include both nutrients, but not a dual management strategy.

Existing data suggest that phytoplankton biomass and community composition in Lake George are regulated by the relative concentrations of nitrogen and phosphorus. Plots of nitrogen:phosphorus ratios as a function of potential controlling variables such as discharge rates, residence time, and water age could be useful for assessing potential impacts, especially as related to nitrogen-fixing cyanobacterial blooms. Regressions of chlorophyll levels versus nutrient ratios may provide useful empirical functions for assessing responses to changes in loading and/or concentrations. However, in many cases significant correlations between nutrient concentrations and phytoplankton biomass can be difficult to obtain because the ambient dissolved nutrient concentrations in the water column reflect only the residual nutrients—the nutrients not already assimilated by phytoplankton. In this sense, inorganic nutrient concentrations indicate what is available for future growth and may not be related to phytoplankton standing stock at the time of measurement. For this reason, nutrient loading estimates, in addition to measurements of concentrations, are essential for developing realistic budgets and predictions of system responses.

Water Age

The "water age" approach described in the Phase I report may be useful in dealing with nutrient-biomass problems if it can include quantitative characterization of phytoplankton growth responses to ambient nutrient concentrations as the water becomes older. Water age, however, needs to be defined more clearly in the Phase II effort, and a clearer explanation should be provided of how it helps to address the project objectives. Can confidence intervals be calculated for water age to provide some measure of the variability in this parameter? Will water age be location-specific? A graph of water age versus discharge for the main reaches of the river may provide insights into how plankton will respond to water withdrawal scenarios and reduced river discharge.

Proposed Phase II Studies

A clear description was not provided in the Phase II work plan concerning how two proposed activities are related to the overall project goals. First, the study of the effects of DOC on the inhibition of microbial activities (Appendix C) seems experimental and peripheral to assessing the potential impacts of water withdrawal. Changes in the loadings and composition of DOC should be determined prior to experimental assessments of its potential effects on the activity of microbial communities, in order to provide guidance and a clear justification for the proposed studies. The dinoflagellate toxin studies (both the cyst and animal surveys) are exploratory and do not offer a mechanistic (either theoretical or empirical) justification that can be linked to water withdrawal scenarios. Second, mesocosm studies to assess the effects of cyanobacteria on zooplankton grazing are also proposed in Phase II. These studies could provide useful insights into the effects of cyanobacterial blooms on trophodynamics in affected waters. However, the mesocosm seem too small and the incubation period too long. Artifacts increase with the duration of such experiments because of algal growth on the walls of the tank, nutrient depletion, mixing conditions, and other factors (Carpenter, 1996; Chen et al., 2000; Porter et al., 2004). Scaling up results from the mesocosms to the river system will be complicated by these potential artifacts. All of these factors must be carefully considered before proceeding with mesocosm studies.

The workgroup is encouraged to continue the development and validation of three-dimensional water quality simulation models (e.g., CE-QUAL-ICM) for the major reaches and lakes of the St. Johns River. These models will be invaluable for scenario simulations.

Summary

The key issues for the plankton, nutrients, and TMDLs workgroup to consider in Phase II involve (1) determination of nutrient and CDOM loading estimates for segments of the middle and lower St. Johns River under the various water withdrawal scenarios, (2) tighter integration with the hydrodynamics and biogeochemistry groups, (3) consideration of potential impacts on bacterioplankton, and (4) implementation of dynamic simulation modeling (CE-QUAL-ICM). Furthermore, the proposed mesocosm experiments could provide useful results, but potential problems with artifacts and scaling need to be considered.

BENTHOS

Benthic macroinvertebrates live in or on the bottom substrate of aquatic environments and have been used for decades to understand the ecological dynamics of streams, wetlands, and coastal marine environments as well as to study the effects of human activity on aquatic ecosystems. Knowledge of these facts led the District to define a benthic workgroup as a crucial program within the WSIS. Both freshwater and brackish waters were to be included in their analyses. The first phase of the effort emphasized three major activities:

1. Explore the state of knowledge on the ecology of benthic environments, including both the organisms themselves and aspects of physical environments upon which these species are dependent. Contracts with several scientific leaders in the field facilitated the efforts of this workgroup.
2. Develop a logical approach to the study of benthic macroinvertebrates as a way to understand the biological consequences of changes in flow or related factors caused by water withdrawal, and
3. Examine available data for insight about the effects of hydrologic alterations from water withdrawals on benthic macroinvertebrates.

The Phase I report reviewed past work in the watershed, began analysis of an existing data base, and briefly described several conceptual models to guide thinking about how to study and understand the effects of water withdrawal on the trophic organization of benthic invertebrates. The conceptual models also described how changes in water level might influence trophic levels above and below the dominantly plant-feeding benthic invertebrates. In addition to a focus on trophic organization, the Phase I report also targets a few invertebrate taxa (e.g., crayfish, apple snail, blue crab, and penaeid shrimp) considered to be of special interest to the WSIS.

Assessment of Phase I

Benthic macroinvertebrates are used as indicators of habitat condition, including the effects of hydrologic alteration on habitat, and more generally to assess the biological integrity of aquatic ecosystems. Biological integrity, an explicitly framed objective of water resource management under the Clean Water Act (Section 101(a)), refers to the condition or character of living systems in the relative absence of modern human activity.

The authors of the Phase I report noted that, at least in general terms, much is known about the effects of water withdrawal on freshwater macroinvertebrates in the St. Johns River. However, little detail was provided on the lessons of the studies cited in Appendix 1 of the Phase I report to the St. Johns River situation. For example, how are invertebrate assemblages changed (e.g., shifts in taxa richness, composition, density, and trophic organization) as a result of water withdrawal? What papers document and demonstrate those or other patterns? Are the situations (benthic macroinvertebrate species composition; waterbody type and size; environment type, biogeographic context) for those papers similar to those in the St. John River? Without these details, it is not possible for the committee to evaluate the relevance of the assertion that much is known about the effects of water withdrawal on freshwater macroinvertebrates to the situation in the St. Johns River. Those kinds of insights and interpretation need to be clear in the body of the report, rather than buried in an extensive, unsynthesized appendix tabulation.

One subject not touched in the Phase I report is the effect of water withdrawal on macroinvertebrates with meroplanktonic larval stages, a group of taxa with high relevance in a comprehensive ecological assessment. The potential impacts of water withdrawals for these early life history stages were not addressed adequately in the Phase I report, a subject that might be addressed jointly with the plankton workgroup during Phase II. Questions that could be address include: what proportion of the species in the St. Johns system have meroplanktonic larvae in freshwater or estuarine regions? How much emphasis needs to be placed on study of meroplanktonic larvae in those two environment types?

An important insight acknowledged in the Phase I report is the need to explore existing data. One such data set was explored but the results were not very robust and, in the committee's view, cannot be generalized without analysis of more detailed data. The need to collect more data is recognized in the Phase I report and a first level plan is fleshed out in the Phase II work plan. A sampling design for the upper reaches of the river is described briefly on the Phase II work plan (SJRWMD, 2009c, pages 54-55) as a first effort to move forward in this important area.

In contrast to freshwater areas, the Phase II work plan concludes that no new data will be collected in the estuarine segments of the watershed. Instead, data from existing studies will be further analyzed in 2009/2010 with the goal of evaluating the effect of salinity and other water quality variables on invertebrate community structure. Without more information on the context of those historical data collections (e.g., timing, spatial distribution, duration, kinds of data, data collection protocols), it is impossible to judge the merit of that decision. Rather, it would be better to defer any decision about additional data collection from the estuary until a careful evaluation of the utility of existing data is completed. Other factors—such as new dredging by the port authority or the Navy—that might affect salinity, sedimentation, or other conditions in the lower river may also require reevaluation of estuarine monitoring decisions.

Recommendations for Phase II

Several of the WSIS workgroups described in general terms the need to identify indicators that can provide reliable and easily interpreted signals about the condition of sites influenced by water withdrawals. Often, those descriptions focused on individual species as indicators. The selection of species for special study seems logical.

A second approach for monitoring benthic invertebrates is the study of trophic groups. This promising approach is warranted, and the outside experts selected by the district are the leaders in this field. The committee strongly supports the development of indicators beyond individual species in the WSIS, and commends the benthos team's commitment to use of more integrative measures of biological condition to judge the effects of water withdrawals.

In addition to the study of functional or trophic groups, the Phase II work plan suggests the exploration of a broader array of measures to characterize biological condition (SJRWMD, 2009c, page 58). Examples of such measures include taxa richness or relative abundance of tolerant and intolerant taxa, taxa richness of selected taxonomic groups such as mayflies or caddisflies, taxa richness of ecological groups such as clingers, and dominance by a few taxa. Although the committee endorses a Phase II effort to use such broad measures, it notes that sufficient detail was not provided on how relevant measures will be defined and used. It is not clear how this activity will be guided by the large body of work in this area developed over the past three decades (see Table 4-1 for selected examples relevant to this project from the literature).

There is precedent for the use of multiple biological metrics as indicators to understand the impact of water withdrawals. For example, a recent study (Freeman and Marcinek, 2006) evaluated fishes in 28 streams used for municipal water supply in the Piedmont region of Georgia. Increasing the withdrawal rate increased the odds that a site's Index of Biotic Integrity score would fall below a regulatory threshold indicating biological impairment. Estimates of reservoir and withdrawal effects on stream biota revealed by such broad measures can be used in predictive landscape models to support adaptive water supply planning intended to meet societal needs while conserving biological resources.

Summary

The committee supports the District's commitment to study of the effects of water withdrawal on benthic invertebrates. Both the Phase I report and the Phase II work plan demonstrate an evolving approach to the study of the effects of water withdrawal on invertebrates. The literature and approaches used by the workgroup should be extended from selected individual species to integrative views of invertebrate assemblages.

To date, the available data are insufficient to precisely define what the effects of water withdrawals on the benthos of the St. Johns River will be. However, the District is commended for sketching the rudiments of a sampling and data analysis program as part of the Phase II work

TABLE 4-1: Selected Literature on the Use of Broad Indicators of Biological Condition

General References	Estuarine and marine systems
Keeler and McLemore 1996	Deegan et al. 1997
Norton et al. 2000	Jordan and Vass 2000
Larsen et al. 2001	Hughes et al. 2002
Fore 2003	Niemi et al. 2004
Yoder and DeShon 2003	Bilkovic et al. 2005
Blocksum 2003	Bradley et al. 2008
Niemi and MacDonald 2004	Fisher et al. 2008
Davies and Jackson 2006	Borja et al. 2008
Karr 2006	**Statistical issues**
Stoddard et al. 2006	Norton et al. 2000
Wardrup et al. 2007	Larsen et al. 2001
Yoder and Barbour 2008	Pont et al. 2009
Benthic macroinvertebrates (freshwater)	**Indicator selection and hierarchies**
Houston et al. 2002	Niemi and MacDonald 2004
Klemm et al. 2002, 2003	Wardrup et al. 2007
Booth et al. 2004	**Effects of hydrological change**
Bonada et al. 2006	Booth et al. 2004
Fore et al. 2007	Freeman and Marcinek 2006

Note: This set of references is not meant to be a comprehensive introduction to what is now a very extensive literature. However, most of the papers cited were published in the last decade, and thus they provide a pathway to the more extensive literature available on the development and use of biological indicators.

plans. Much of the detail of those efforts is yet to be defined, a fact that makes it difficult to offer specific recommendations on the next steps to be taken in the process.

LITTORAL ZONE

The proposed surface water withdrawals from the St. Johns River are likely to exacerbate salinity intrusions in the estuarine portion of the river. This could have detrimental effects on species of submersed aquatic vegetation (SAV) that have adapted to fresh or brackish water, such as wild celery, *Vallisneria americana*. The littoral zone workgroup is assessing the potential damage to submersed vegetation from the proposed water supply withdrawals via an extensive monitoring program. By assessing the condition of the SAV during high salinity pulses in the lower river and comparing their data to literature values on salinity tolerance, they plan to estimate the effects of water supply withdrawals on SAV. Though not explicitly part of its charge, the workgroup is also trying to better understand related impacts on dependent species such as invertebrates and fish and on water quality.

The focus of the Phase 1 effort was on the lower 131 km of the St. Johns estuary, where SAV abundance has been monitored intensively in three separate surveys over the last ten years. Based on aerial photography and field surveys, the lower St. Johns River had approximately 866 hectares of SAV in 2003–2004 (Dobberfuhl, 2007); about 63 percent of the total cover is due exclusively to *V. americana* (Sagan, 2007). This is a perennial species with well-developed underground roots and rhizomes that are ideal for consolidating bottom sediments, providing oxygen to benthos and promoting nitrification. In addition, *V. americana* beds are excellent habitats for small fish. Other plant species found to have significant cover in the St. Johns include *Najas guadalupensis* (16 pecent) and *Ruppia maritima* (10 percent). Both species have less below-ground biomass and are thought not to be as effective in binding or in oxygenating the sediments as *V. americana*.

Assessment of Phase I

The District made an excellent start in the Phase I report and used its extensive monitoring programs to full advantage. The littoral zone working group also developed a salinity exposure model from the literature on *V. americana* that will be useful to predict the effects of increasing salinity on this species in the lower St. Johns River.

Based on the Phase I efforts, the workgroup reached a number of important conclusions. The District predicted that projected future water withdrawals could have dramatic consequences on SAV in some areas, especially where *V. americana* populations now fluctuate in the lower St. Johns River. A large portion of the estuary is sensitive to wet- and dry-year fluctuations, and any hydrological alteration may cause a shift in salinity as the salt wedge moves upstream. Higher salinity could cause a collapse of the *V. americana* in one of the most important segments of the St. Johns River in terms of reducing habitat as well as limiting nitrification and denitrification potential (Wigand et al., 1997). Unfortunately, higher-salinity-tolerant species such as shoal grass, *Halodule wrightii*, or widgeon grass, *R. maritima*, do not now appear to be able to re-vegetate the areas left barren if *V. americana* is lost. Although *V. americana* presumably could migrate further upstream, there is less shallow water area there, and so a net loss of habitat is still expected. To better understand the potential for this, it would help to have a graphic displaying the depth distribution by segment (i.e., the hypsometry) of the system compared to the *V. americana* habitat under varying flow (i.e., salinity) conditions. Hydrologic alterations also may be expected to change patterns of SAV sequestration of phosphorus, as well as diminish the overall amount of nitrification and denitrification if SAV declines in the estuary (Kemp et al., 2005), making the benthos less resilient to nutrient loadings expected from the increase in population.

Hydrodynamic model simulations show that if 262 MGD were withdrawn from the river, average bottom salinity would increase over 0.7 ppt from river miles 17 to 37. Although this is a small shift, it may cause the loss of *V. americana* where it is now marginal with regard to salinity tolerance. Planned channel deepening efforts in the lower St. Johns River could compound this increase in salinity up to another 2.7 ppt (Lowe et al., 2008), and sea-level rise could increase salinity in the same section of the river by as much as another 0.7 ppt (high projection; SJRWMD, 2008). All these factors could cause considerable stress on the existing *V. americana* beds in the lower St. Johns River. It is equivocal whether salinity increases will promote invasive SAV species in the St. Johns River, and in fact the reverse could be the case. That is, if salinity rises significantly in what is now the tidal freshwater portion of the river, the exotic species *Hydrilla verticillata*, which can form dense mats, would actually be much less competitive, allowing native SAV species to flourish.

Recommendations for Phase II

Several suggestions are offered here to improve upon the Phase II data collection and analysis efforts in the littoral zone.

Water Quality Monitoring and Data Analysis

The hydrologic modeling in Phase II is expected to provide more spatially explicit predictions on the salinity increases of the littoral zone. In Phase II, three separate water quality monitoring programs are to be continued that will provide important data for verifying the model output. Submersed species distributions are variable from year to year; thus, comparisons with salinity data collected at multiple locations will be useful in refining the salinity relationships established from literature values.

The District also should add at least one continuous monitoring salinity station in the littoral zone during Phase II. The District currently measures salinity every 3 to 4 days in SAV beds, but continuous monitoring in the shallows could detect short-term salinity excursions where *V. americana* is at risk. These data should be compared to the monitoring data now collected by the U.S. Geological Survey (USGS) in the main channel.

It may be helpful to display water quality data being collected and SAV distribution along the axis of the St. Johns River using the graphical techniques of Stevenson et al. (1993) and Staver et al. (1996) in the Choptank River, which provided the basis for determining critical thresholds for nitrogen, phosphorus, chlorophyll, and suspended sediments that the Chesapeake Bay Program has used for survival of SAV. These approaches, as well as that described by Biber et al. (2008), which would help factor in light attenuation due to CDOM, should be further investigated for their applicability to understanding the impacts of water withdrawal to SAV in the St. Johns River.

Mesocosm Studies

The focus of this workgroup has appropriately been on *V. americana,* and significant progress has been made in defining salinity tolerances of this species from literature values. Unfortunately, recent genomic studies suggest that several strains of *V. americana* exist, and each has different environmental niches (Katia Englehardt, University of Maryland, 2009, personal communication). Consequently, more studies on salinity tolerance of local populations from the St. Johns River would be instructive to validate the values derived from the literature. It would be beneficial if mesocosm studies could be planned to determine whether the salinity–stress–mortality relationships described in Table 1 of the Phase I report (p. 35) are accurate for *V. americana* collected from the St. Johns River. At a minimum, three populations along the salinity gradient should be selected for study. The response of young versus mature plants as well as changing light levels should be examined. The mesocosm studies should be conducted under high and low light conditions that characterize blackwater systems like the St. Johns River. Photosynthetically Active Radiation levels derived from the field program should be used as reference points for the mesocosm studies.

Ecological Assessments   57

Because the withdrawal of 262 MGD from the lower St. Johns River is projected to significantly affect the existing beds of *V. americana*, the District also should assess whether any other existing SAV species might be able to take its place as a dominant macrophyte in the littoral zone. Based on the field data, *R. maritima* might be the most likely candidate (but District staff might suggest others). A mesocosm program similar to the one described above for *V. americana* would help explain why *R. maritima* will not recolonize when salinity is elevated (as is now the case in major East Coast estuaries such as Chesapeake Bay—see Stevenson and Confer, 1978; Fredrette et al., 1990; Batiuk et al., 1992; Stevenson et al., 1993). Both salinity and light responses (especially important in the turbid portions of the lower St. Johns River) would need to be a part of such a mesocosm study. Though these experiments go beyond the planned Phase II work, the committee believes that they would bolster the important monitoring work now being done by the littoral zone workgroup.

Summary

The District is to be commended for its studies to better understand how the proposed surface water withdrawals will affect local SAV populations. The littoral zone workgroup has already made preliminary estimates of the effects of water supply withdrawals on SAV, which will be improved upon during Phase II with more advanced hydrodynamic modeling. To enhance the SAV monitoring program, the District should consider adding at least one continuous salinity monitoring station in the littoral zone during Phase II to detect short-term salinity excursions where *V. americana* is at risk. The workgroup should also undertake more study of salinity tolerance of local populations from the St. Johns River, perhaps via mesocosm studies, in order to validate the values derived from the literature.

Because the workgroup predicts that projected future water withdrawals could have dramatic consequences on SAV in some areas, it would be worthwhile for the workgroup to assess whether any other existing SAV species might be able to take the place of *V. americana* as a dominant macrophyte in the littoral zone, for example *R. maritima*. A mesocosm program similar to the one described above for *V. americana* would be helpful in this regard, although it is acknowledged that such experiments go beyond the planned Phase II work.

FISH

The District's Phase I report describes how water withdrawals could influence spawning success and recruitment of important recreational and commercial fishes, populations and distribution of other fish species, and critical dimensions of fish habitat. In the Phase I report, potential effects of water withdrawals on fishes were divided into two main categories: direct effects from entrainment or impingement, and indirect effects associated with changes in habitat.

The exposure of fishes to potential effects from water intake structures will vary based on their life history characteristics in relation to the location of the intake structures. The District identified 22 species that it considers good "indicators of impact" due to water withdrawals (although the 22 species listed in Table 1 of the Phase I report and discussed in the associated text do not exactly match the species with potential impacts identified in bold in Appendix A). Each of these species has life history characteristics that may make it vulnerable to habitat alteration stemming from reductions in flow or stage, or it has regulatory or ecological importance. This section provides general suggestions related to (1) entrainment and impingement, (2) habitat changes, and (3) other issues as the District continues to refine its assessment of the impacts of water withdrawals on St. Johns River fishes. Unlike the other five workgroups, the recommendations for this workgroup are found throughout the section, which is organized by the three topics mentioned above.

Entrainment and Impingement

The District recognizes the importance of understanding the potential effects of entrainment and impingement from surface water withdrawals. Its hypothesis is that given the relatively low intake velocities of the proposed withdrawals, adult and juvenile fish entrainment or impingement probably will not be significant enough to elicit broad-based community changes. However, fish eggs and larvae lack the mobility to avoid being entrained by these water intake structures. Because existing data are limited, the District has contracted a study of larval fishes in five regional locations in the upper and middle basins of the St. Johns River. The sampling will be located near the proposed surface water intake structures (i.e., near State Road 50, Lake Jessup, Yankee Lake) and within the center of what is reported to be the primary spawning grounds of American shad (*Alosa sapidissima*; Family Clupeidae, the term "herrings" afterwards includes all family members) in the St. Johns River.

The sampling schedule outlined in the Phase I report is unclear as presented, but our interpretation is that a sampling event will occur once every three to six days (sampling at all sites) during the February to May herring spawning season and once every seven to 12 days for the rest of the year. According to the Phase I report, on each sampling event two ten-minute tows at six sites within each of four regional locations will be collected (48 total collections per sampling event). Closer examination of Appendix B in the Phase I plan, however, reveals the potential of 54 samples taken (27 x 2 tows = 54 total samples). According to the recently released work plan, Phase II changes the plan from four to six regions with a total of over 6,000 larval samples to be collected, but the details of where the additional sampling will be and its rationale are not yet available. It will be important for the District to clearly define the actual distribution and frequency of Phase II sampling especially as potential water withdrawal locations become more concrete.

The committee identified several issues of concern about the larval fish study in the Phase I report, including the frequency of larval fish sampling in these areas, the lack of nocturnal sampling, and the narrow focus on clupeid larvae. First, larval fish density in marine, estuarine, and freshwater environments is always highly variable on small and large spatial and temporal scales including day and night differences (Houde and Lovdal, 1985; Shaw et al., 1988; Allen and Barker, 1990; Peterson and VanderKooy, 1995; Fowler-Walker et al., 2005). These larval density estimates are likely to vary whether the District samples on day 3 or day 6 within one sampling period as much as between the sets of 3- to 6-day periods or the 7- to 12-day sets. Thus, density estimates may be high or low over short time periods as well as spatially within the system studied.

Second, the results could be somewhat compromised because many species are more active at night or are more easily captured at night (Shaw et al., 1988; Morse, 1989; Gray, 1996). Although the Phase II day and night sampling at Lake Monroe (focused on herring) will be useful, this may bias the density estimates used to evaluate the effects of the water intake structures elsewhere. While there are substantial costs and logistical issues associated with increasing the night sampling, night sampling in at least one other area up- or downstream of Lake Monroe would allow the District to capture larval fishes and patterns of impact that would otherwise be missed.

Finally, focusing the sampling on those periods most relevant to herring (February to May with a more relaxed sampling schedule the rest of the year) could be problematic. Many other species are important in the maintenance of a sustainable and diverse fish fauna in the St. Johns River system in the face of expanded water withdrawals and should be carefully examined. The committee suspects that for some species of interest more frequent sampling will be needed because of the extremely high variability inherent in larval fish sampling protocols, particularly for those species that are not abundant. At this juncture, a standard power analysis (Sokal and Rohlf, 1995) needs to be conducted on the larval fish data that have been collected to date to see if the sampling is adequate to address the intended goals for the species of interest.

In the Phase II work plan, the Empirical Transport Models (impingement and entrainment quantification procedures) appear reasonable, but the District needs to provide references to those models that show that the assumptions of the approach (egg and larvae are uniformly distributed across the channel) are reasonable or even attainable and that also speak to their accuracy in previous studies. Other needs for providing clarity about these models (and thus their potential success) include the following. First, for which species will the District estimate density and thus potential IL (defined as the total number of individuals of a species lost to entrainment)? Are length vs. swimming speed data available in the literature? Second, if such literature data are not available (that is, not at all or not for many size stages), then the District needs to outline what will be done. Finally, to effectively address loss of juveniles or adults at intake sites, the District must calculate mortality rates of larvae *through* adult fish, which is difficult. The Phase II work plan indicates that the literature will be used for adult fish mortality rates (assuming they are available), but how data for juveniles will be obtained (e.g., otoliths, size-based estimates, or the literature) is unclear. The choice of technique and the availability of literature data are vital to the model outcomes, and more detail needs to be provided.

Habitat Changes

In addition to impingement and entrainment, the Phase I report describes other potential environmental impacts to fishes as a result of water withdrawal due to potential habitat alteration, particularly with regard to salinity and water level changes. Such indirect impacts can be considerably more difficult to assess, especially as scientists have begun to better understand the complexity of these systems. For example, nursery habitat is no longer defined simply based on easily measured structural features associated with feeding opportunities and protection from predators, like marsh edge, open water, SAV, and oysters. Instead, a more nuanced perspective, as well as spatially and temporally dynamic view of estuarine environments and landscapes, is becoming apparent, in which habitat diversity, quality, and quantity are all nested within a framework of abiotic variability (Simenstad et al., 2000; Peterson, 2003; DeLong and Collie, 2004; Peterson et al., 2007).

Salinity

Water withdrawals in the St. Johns River could affect the extent of saltwater intrusion in the downstream portion of the river, impacting the distribution of both estuarine and freshwater fish communities. According to the Phase I report, the estuarine portion of the lower St. Johns River is "an extremely variable area that can experience wide fluctuations in salinity and other environmental parameters over tidal and seasonal cycles, and in response to strong winds and rain events." The local and regional fish fauna are certainly highly adapted to survive and even thrive in this historically variable environment. However, it would be a mistake to conclude that altered patterns of variability in salinity caused by water withdrawal would have low or low–moderate impacts directly on larval and juvenile marine, estuarine, and freshwater fishes in the region. This is because the *rate* of salinity change, which may be altered by withdrawals, strongly influences community structure in fishes along tidal gradients (Peterson, 1988; Peterson and Meador, 1994). Moreover, fishes can be affected via salinity impacts to (1) SAV, which is vital to many juvenile fishes, and (2) predator–prey dynamics along the axis of the St. Johns River.

The District notes that species abundance, distribution, and dispersal patterns of larval and juvenile fishes in the St. Johns River estuary are not well documented, such that the "likelihood of impact" predictions to estuarine fishes in the Phase I report are based on older data for selected adults (Walburg, 1960; Moody, 1961) or on anecdotal data. Indeed, the information provided in the Phase I report did not include data on, nor consider the effects of water withdrawals on, fish assemblages as a whole, fish spawning, and most life stages (not just larval fish stages) that use the lower St. Johns River as a nursery. The District notes that more detailed information on factors affecting the abundance, distribution, and recruitment success of important estuarine species would be useful and points to a long-term state juvenile fish monitoring program started in 2001 near the mouth of the St. Johns River (FFWCC, 2001) and extended upstream in 2005 (Solomon and Brodie, 2007).

The Phase II work plan notes that the Florida Fish and Wildlife Conservation Commission (FFWCC) has been placed under contract to continue the studies mentioned above in the lower St. Johns River (up through Palatka) and to examine the influences of water level and salinity changes on fish and selected decapod assemblage structure. The District's development of a contract with FFWCC to analyze their long-term juvenile fish survey database is a step in the right direction that should improve the Phase II study. Analysis of these data sets should focus on how assemblages of fishes and selected decapods (those routinely recorded by the state Fisheries Independent Monitoring [FIM]) would respond to water level changes and particularly the extent of expansion or contraction of vital nursery areas relative to saltwater intrusion. It should be noted that the data collection approaches used by the state may miss important resident taxa (e.g., *Fundulus* spp., *Cyprinodon variagatus*, *Lucania parva*, etc.) not susceptible to small trawls and large seines. Furthermore, the FIM sampling may not be spatially and temporally defined enough to capture the data needed to address water withdrawals and subsequent movement of the saltwater upstream. These concerns, which relate to where the data are collected and how those sites relate to changing patterns of salt intrusion, water level fluctuations (particularly under low flow conditions), and the saltwater–freshwater transition zone, should be kept in mind as the District analyzes the FFWCC data.

The spatial and temporal variation in salinity noted by the District in the lower St. Johns River leads the committee to question how the District plans to manage that specific range of variability in the future, when the effects of water withdrawals may be coupled with additional river dredging by the port authority and the Navy to allow ship access further upstream. Managing for any range of variability is not the same as managing for the range within which these species have evolved. If the range of variability changes, it may influence future year-class strength of some species and assemblage structure in these same geographic areas along the transition zones from saltwater to freshwater. Shifting the saltwater–freshwater transition zone to different areas may also influence normal migratory routes and timing for some taxa (fish and other nekton) with negative effects on survival during various life-cycle stages.

Finally, the fish data analyses need to be integrated with other components of the regional biota (e.g., benthos, decapods, SAV)—factors that also may be influenced by salinity shifts. The cumulative interaction of the assemblage will be critical to understanding fish responses to salinity (Peterson, 2003) and thus to water withdrawals. Indeed, a program could be developed that integrates the FFWCC fish and decapod data and the benthic macroinvertebrate and aquatic vegetation components using an approach similar to that of Peebles and colleagues from the University of South Florida. Peebles et al. (2007) developed an approach for evaluating the effects of water withdrawals on the egg, larvae, juvenile, and adult stages of the bay anchovy, *Anchoa mitchilli*, in a number of Florida estuaries. This approach could be modified to incorporate taxa (beyond fish), using data from the WSIS and FFWCC projects. Such an approach would allow a more realistic examination of how much the nursery areas of the Lower St. Johns River might expand, degrade, or be lost due to saltwater intrusion (e.g., Kimmerer et al. 2009).

Water Level Changes in Transition Zones and Floodplains

A major concern with the Phase I report is the lack of consideration of water level decreases (projected to be as much as 4 cm in the upper St. Johns River basin) and the influence that those water level changes will have on fish population dynamics and distribution. Many important factors for fish survival from spawning habitat to foraging opportunities and the availability of refuges may be influenced by water level declines in complicated ways. Indeed, much is known about shallow vegetated areas in lake and stream systems that are important fish habitats and that drive life history parameters such as reproduction, growth, foraging, and distribution of fish biomass by habitat type (Keast et al., 1978; Keast and Eadie, 1984; Keast, 1985). Many but not all of these life history parameters are strongly linked to emergent, floating, and submersed aquatic vegetation.

Centrarchid fishes (sunfishes and black bass) provide excellent opportunities to understand the dynamics of these relationships and dependencies. This diverse family in the St. Johns River system includes many valuable sport fish throughout their range. Aquatic vegetation mediates numerous ecological processes in aquatic habitats, specifically reproduction, foraging and predator–prey interactions (Theel et al., 2007; Kovelenko et al., 2009). Aquatic vegetation could be negatively affected by water level reductions in the shallow and relatively flat floodplains of the middle and upper St. Johns River. These studies also noted that many of the impacts to trophic pathways of sunfishes, beyond direct changes in interstitial space in aquatic vegetation, are indirect due to changes in macroinvertebrate assemblages and substrate heterogeneity (Kovelenko et al., 2009).

Given the spatial variation in geomorphology and microtopography of the floodplains along the St. Johns River, important habitat characteristics are likely to be modified by a 4-cm water level drop spread across a large area. Therefore, the District should carefully study potential impacts to all fishes in the middle and upper St. Johns River, but particularly those that require shallow areas for spawning and foraging, including but not limited to the centrarchids (Breder and Rosen, 1966). The District should also develop a comprehensive study of spawning habitat in the upper St. Johns River relative to the projected 4-cm water level drop and use GIS to evaluate how much area will actually be lost in terms of drying out or converting to upland habitat.

In the Phase II work plan the District has made great progress explaining its approach to examining how water level changes might impact floodplain habitat for fishes. It will develop a detailed literature review on fish use of floodplains in Florida and then couple those literature patterns with selected minimum flow and level (MFL) profiles in the St. Johns River for which it has calculated the loss of floodplain habitat based on predictions of water level change. The District suggests in its Phase II work plan that it will conduct these habitat availability studies in areas being considered for consumptive use withdrawals during months when reproduction of a number of fish species is documented to occur (although details of this "documentation" are not given). This is reasonable, but the committee questions how the District will go from change in available floodplain area for a given water level change to the fish production metrics that are vital to sustainability. Some fish species use floodplains to actually spawn, while the adults of other species use it to forage prior to spawning—a very big difference that may not be captured by the proposed analysis.

Impacts to Anadromous fish

In the Phase II work plan, the District added an ongoing cooperative (University of Florida, FFWCC, and the District) component to their collection activities addressing tracking juvenile and adult herring, as the populations of three herring species are depressed. The goal of this activity is to determine migratory and spawning habitat of herrings, and then to compare the needed habitat characteristics with what is actually available in the St. Johns River before and after potential water withdrawals. This approach appears reasonable, but few details are provided to allow for a critical evaluation; if possible, it would be appropriate to link these data with the larval herring data sets for a more complete picture.

Other Issues

In the Phase I report, the District categorized fishes as commercial, recreational, or forage fishes. Although some fish species are commercial or recreational (or both), this somewhat arbitrary classification narrowly views the remaining species in the St. Johns River as important only as food for the commercial and recreational species. The District should take a broader view in Phase II. Dozens of other fish species are permanent residents or seasonal visitors to the St. Johns River. An understanding of their ecologies and how their distributions and abundances will be affected by proposed water withdrawals is central to the evaluation of natural resource effects that comprises the bulk of the WSIS. For example, depending on exactly where the water intake structures will be located and how many structures will be built, the narrowly framed forage category may be the most affected group. A number of other fish species listed in Appendix A of the Phase I Report, such as two sturgeon species (shortnose and Atlantic), rare and habitat-specific sunfishes (3 *Enneacanthus* spp.), rare pygmy sunfish (Everglades and Okefenokee), and the River goby (*Awaous banana*) could be sentinels to any withdrawal project as they may be some of the first species to be affected.

In the Phase I report, the District developed a graded impact score approach to judging the "likelihood of impact" (e.g., low, moderate, and high) from water withdrawal. The report, however, does not provide any information about the origin and foundation of those scores or the nature of the background criteria used to judge these impacts, other than to say that the "likelihood of impact" is based upon the District's analysis of the risk compared to the preliminary hydrologic modeling results. Are these scores based on some quantitative or semi-quantitative analysis or are they a reflection of expert opinion? A description of the process underlying these "likelihood of impact" scores is needed to understand the predicted outcomes. Of the 15 potential impacts related to all categories outlined in Phase I (freshwater influences, water quality, river-floodplain interactions, and estuarine influences), only one "likelihood of impact" is classified as moderate-to-high (current reversals) and one as moderate (effects on winter spawning migrations of the American shad); the projected likelihood of the remaining factors are low-moderate or low. Can these scores be justified given the lack of solid data (including spatial and temporal patterns) on fishes of the St. Johns River system? For example, a considerable number of references are based on studies of clupeids in reservoirs and lakes, for which the characteristics may vastly differ from those of riverine fish.

Of the 22 species listed in Table 1 (pages 11-12 of the fish section in the Phase 1 report), the "likelihood of impact" for 13 of the species is categorized as "low," "negligible" for four, "low-moderate" for one, and "moderate-to-high" for one. Overall, the Phase I report concludes that almost none of the identified fish will be significantly affected either directly via entrainment and impingement or indirectly through habitat loss or degradation and salinity intrusion. The committee is not convinced that the information available justifies this interpretation, and it looks forward to the more complete risk analysis that is slated to be part of Phase II.

Summary

A number of decisions need to be made by the District that are crucial to the development and clarity of a final fish program and study design. Based on the Phase I report, the District should evaluate the merit of at least three additional program activities before final sampling protocols are developed: (1) analyze the larval fish data collected to date to determine if the sampling protocols are adequate to address the intended goals, (2) integrate existing fish data sets with benthic macroinvertebrate and SAV data to address impacts to fish assemblages, and (3) comprehensively study the impact of a 4-cm water level drop to the spawning and feeding habitat in the middle and upper St. Johns River.

In the Phase II work plan, the District did not address adequately the first item mentioned above, but it did enhance and clarify the second and third items. The second item will now be addressed, in terms of fishes and selected decapods, by the FFWC FIM program contract now in place. It should be kept in mind that while the FIM program is a great enhancement to the overall WSIS, it does not collect samples in small creeks in the lower St. Johns River and thus will miss important nursery areas. In terms of the third item, the District needs to elaborate on how the loss of available floodplain area will translate into loss of fish production, which is vital to a more complete understanding of the effects of water level changes.

WETLANDS AND WETLAND-DEPENDENT SPECIES

Changes in hydrology can alter the structure and function of wetlands. Therefore, the wetlands workgroup is examining the potential impacts of the proposed surface water withdrawals to wetland vegetation and a few species of wetland-dependent fauna. The workgroup also plans to address indirect effects resulting from changes in water quality following withdrawal, including the value of wetlands as habitat for wetland-dependent species. The District predicts that impacts could range from changes in vegetation community type or structure (including species composition), altered productivity, and shifts in the position of boundaries between communities. Although hydrologic alterations are expected to impact a broad array of wetland functions, only the habitat function was addressed in this section. In general, the Phase I wetlands work is descriptive and conceptual in nature. It is planned that the conceptual model developed in Phase I will be developed into a fully operational GIS-based model in Phase II of the study.

Assessment of Phase I Report

This section includes an assessment of the Phase I wetlands work along with some related recommendations for improvements in the proposed Phase II. Issues discussed include the methods used to identify floodplain wetlands, the GIS-based determination of sensitive wetland areas, the use of wetland-sensitive fauna as indicators of hydrologic change, and maximizing the use of available data.

Identification of Floodplain Wetlands

The WSIS is limited to wetlands located on the floodplain of the St. Johns River (i.e., wetlands directly influenced by the river's hydrology). A GIS analysis was conducted in Phase I to define the floodplain and identify wetlands that lie within it. With a few exceptions,[1] the floodplain was delineated using the 5-foot contour line (the line representing points of equal elevation at 5-ft above sea level) on 7.5-minute USGS quadrangle maps, with the assumption that the 5-ft elevation captures much of the 50-year floodplain. The District considered this

---

[1] At the river mouth, where the 5-ft contour line was not available, a digital elevation model was used to estimate the 5-ft contour, which was then checked for accuracy using aerial photography and wetland maps. At the southern end of the basin, first the 10-ft and then, at higher elevations, the 15-ft contours were used. Where these contours did not coincide with the upland edge of the floodplain, the report states that the floodplain boundaries were drawn in manually.

elevation to overestimate the portion of the floodplain likely to be affected by the river, although no data were provided to support this conclusion.

Given the current limited availability of GIS data for wetland mapping, the District had little choice but to use the relatively coarse topographic data. However, this approach lacks the resolution necessary to predict the ecological effects of hydrologic change. For instance, the floodplain, particularly in the lower segments, is relatively flat with microtopographical features that create and maintain diversity as hydrological conditions (e.g., water levels, frequency and duration of inundation) vary. Fine-scale elevation data in the form of a digital elevation model (DEM) are needed to produce accurate maps in flat, wetland-rich areas, particularly in efforts to characterize hydrology and map wetland diversity (Maxa and Bolstad, 2009). The District has recognized the need for a DEM in order to adequately evaluate wetland response to the proposed withdrawals, and it plans during Phase II to produce DEMs for the portions of the watershed where appropriate data are available.

Recent advances in wetland mapping are available to update maps and improve their accuracy and are recommended for use in this study. One of the best options is the use of LIDAR (Light Detection and Ranging) imagery, which, if used, could greatly increase the sensitivity of models to assess the effects of hydrologic change. LIDAR provides elevation data with high resolution (15 cm to 1 m elevation) and accuracy and can improve efforts to map vegetation structure (Hyde et al., 2006). This also will help characterize changes due to altered hydroperiod, in which small changes can lead to large alterations in wetland function and extent. Thus, the use of LIDAR imagery would allow the systematic monitoring and assessment of hydrologic changes (Lang and McCarty, 2008). According to the Phase II work plans, the District is planning to test the use of LIDAR imagery using data obtained from several counties (see Figure 4-1). Much of the lower and middle St. Johns River floodplain either has LIDAR data available or it is pending. This should provide insight into the usefulness of LIDAR in the District's GIS analysis and DEM development. The District is encouraged to acquire LIDAR imagery for the entire study area as soon as possible, understanding the constraints imposed by limited resources.

Use of previously collected data on plant communities and elevation available from transects sampled as part of the MFL study (as described in Neubauer et al., 2008)—but not included in the Phase I report—also may provide ground truth information for the existing wetland maps and a means to determine their degree of accuracy. Where transect data are available, they can help characterize the extent of hydrologic connectedness (river to floodplain) and identify the locations where impacts from withdrawals are expected to occur. These data can serve as a baseline against which future conditions are measured. The Phase II work plan indicates that the transect data will be used to accomplish this through appropriate data analysis.

Identifying Sensitive Wetland Areas

To cope with the distribution of wetland types found in the watershed, the wetlands workgroup divided the river into nine segments (see Figure 3-3) deemed relatively homogeneous in terms of soils, vegetation, hydrology, water quality, and fauna. For each segment, information was provided on channel length, hydrologic features, salinity, predominant wetland plant species, shoreline ratio, soils, wetland-dependent fauna, and the relative likelihood of impacts

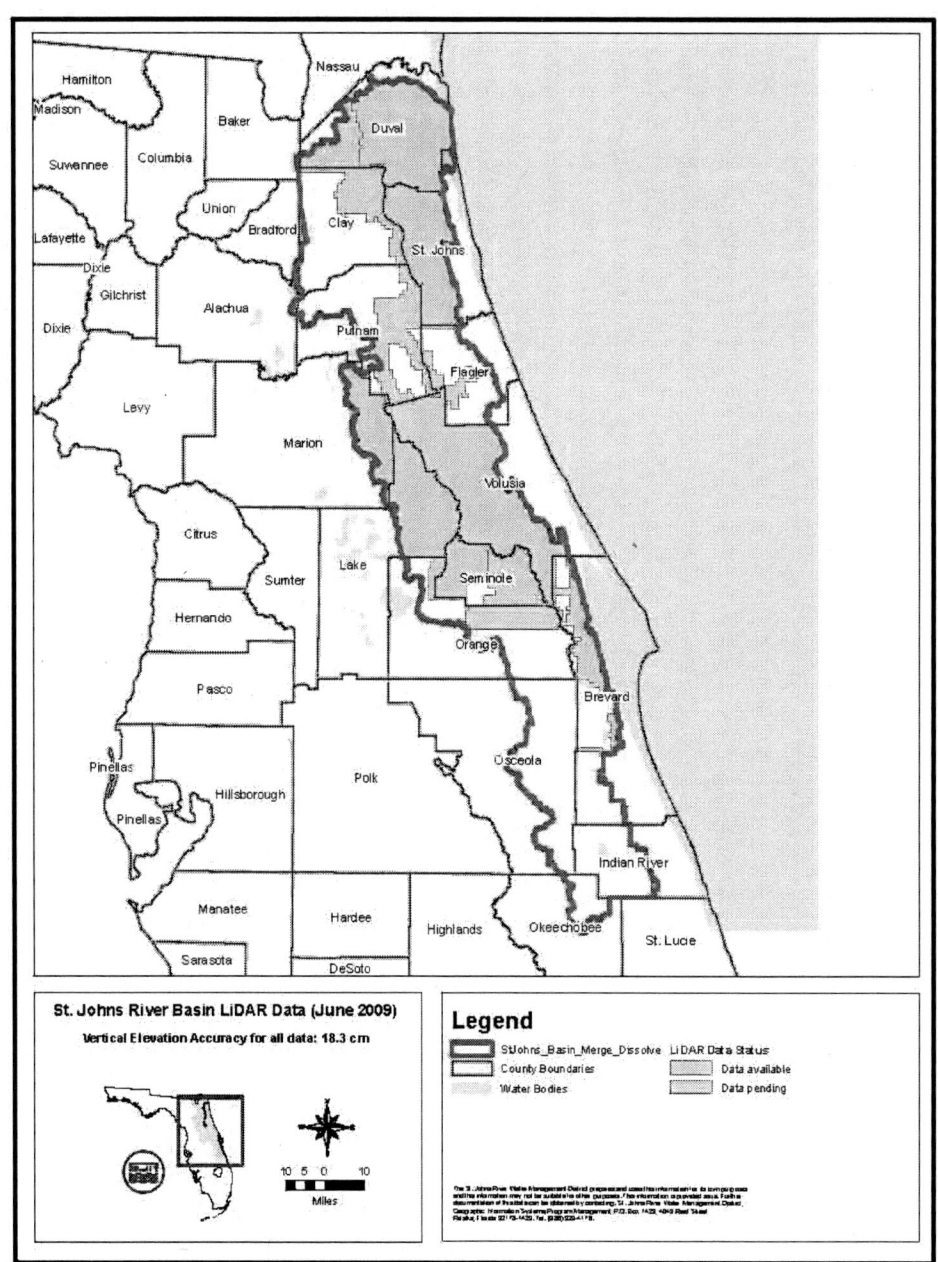

FIGURE 4-1: St. Johns River Basin LIDAR Data (June 2009).
SOURCE: SJRWMD.

from water withdrawals to all wetlands in the segment. This approach is analogous to the River Continuum Concept used to model the relatively predictable changes assumed to occur in river ecosystems from source to mouth (Vannote et al., 1980), and it is an excellent start to address the expected physical and biological diversity that exists within the basin.

However, this conceptual, qualitative approach made no effort to discriminate the varying response of different wetland types within the river segments. Wetlands vary widely in terms of hydrology, landscape position, and the functions they provide. Adopting a classification scheme, such as the hydrogeomorphic (HGM) approach (Brinson 1993), which the District now plans to

Ecological Assessments 67

employ as part of the Phase II work, has been shown effective in stratifying the diversity in species composition and ecosystem function. In conjunction with a more detailed and accurate DEM, this might help the District better cope with the landscape-scale diversity of wetland types and their hydrologic drivers, and it could replace the notion of the "landscape hydrological milieu" used in Phase I. The HGM approach is well established in wetlands research and monitoring programs and has been shown to be effective in reducing wetland variability in order to evaluate the relative response of wetland classes to hydrologic change (Rheinhardt et al., 1997; Whigham, 1999; Wardrop et al., 2007).

The District used GIS data layers along with best professional judgment to evaluate potential wetland response to hydrologic alterations and to categorize the nine river segments according to the "likelihood of hydrologic impacts" (i.e., changing stage, salinity). The "likelihood of impacts" was classified qualitatively (i.e., high, moderate, low, none) as follows:

- Weights were assigned to the various GIS data layers (e.g., wetlands, soils, topography, hydrology) to indicate their relative importance to wetland response.
- A stressor layer was created to represent water level lowering or other environmental changes, including changes in salinity.
- Where sensitive areas overlapped with areas impacted by stressors, some likelihood of effects was assigned.

The committee is concerned that the criteria used to determine the likelihood of impacts are not provided in the report. Specifically, the weights assigned to the GIS data layers were not presented in the Phase I report, nor was there an explanation of how the weights were developed. No information was presented about how the stressor layer was created, nor was a comprehensive list provided of the stressors considered. The committee appreciates the difficulty of this type of modeling effort, but without a description of the weights assigned to the different types of data, details on the stressors included in the GIS database, and the criteria used to estimate the likelihood of impacts, it is impossible to judge the results.

Results of the Phase I analysis are summarized in Box 4-1. This approach was developed and used as a coarse screen to identify river segments that warrant the most attention, which is logical. In large part it is meant to identify wetlands on highly permeable, highly organic soils that are hydrologically linked to the river and dominated by vegetation that is likely to be sensitive to drawdown. However, it is difficult to assess the reliability of this analysis without a detailed description of the rationale, data, and procedures. Note that many of the conclusions listed in Box 4-1 go beyond impacts to wetlands, but rather pertain to the other workgroups.

Wetland-Dependent Fauna as Indicators

The wetlands workgroup presented a literature review of four bird species (wading birds and raptors) known for their sensitivity to hydrologic change that may be useful as indicators of wetland impacts: the wood stork (*Mycteria americana*), white ibis (*Eudocimus albus*), limpkin (*Aramus guarauna)*, and snail kite (*Rostrhamus sociabilis*). The wood stork and the snail kite are federally endangered species, and their water-level requirements for successful foraging and nesting are well known. The white ibis and limpkin are listed as Florida species of special concern. The report lists specific hydrologic ranges required for foraging and nesting of these

> **Box 4-1**
> **Preliminary Assessment of Impacts along Nine Segments of the St. Johns River**
>
> The Phase I report includes a broad assessment of potential impacts in nine river-wetland segments (see Figure 3-3):
>
> 1) **Mayport to Fuller Warren Bridge (River Kilometer [RK] 0 – 39.6)**: this reach is dominated by high-energy tidal exchange. It is unlikely that there will be impacts due to water withdrawals.
> 2) **Fuller Warren Bridge to Flemming Island (RK 39.6 - 65)**: This area could suffer moderately from salinity, although stage effects are likely to be low. Altered salinity may impact SAV in this area.
> 3) **Flemming Island to Little Lake George (RK 65 - 163.1)**: Few impacts due to water level changes are anticipated, but there may be impacts due to altered salinity. The likelihood of effects due to altered salinity is judged to be moderate.
> 4) **Little Lake George to Astor (RK 163.1 - 204.3)**: This area is the focus of the plankton workgroup because of the increased probability of algal blooms. The District reports that wetland effects are likely to be minimal, in part because the area already has high salt inputs from springs.
> 5) **Astor to Wekiva River (RK 204.3 - 253.7)**: Wetlands in this area are dominated by hardwood swamps. This is the first segment above tidal influence, so it is predicted that salinity will not be affected (unless this reach becomes tidally influenced). Stage changes with water withdrawal could be an issue. This segment concerns the biogeochemistry group because of possible changes in the duration of inundation for wetland soils. Small changes in stage may lead to the dewatering of large floodplain areas, and the likelihood of effects due to altered stage is judged to be moderate.
> 6) **Wekiva River to St. Road 46 (RK253.7 - 310)**: This segment is part of the middle river and includes Lakes Monroe, Jessup and Harney. Stage and salinity impacts are likely to be low to low-moderate. The area is important for shad spawning and any potential for impacts to fish communities are of concern.
> 7) **St Road 46 to St Road 520 (RK 310 – 378)**: Segment 7 is predominantly wet prairie and includes Puzzle Lake. There are no tidal effects in this segment, so the potential for stage effects are relatively high (the likelihood of effects due to altered stage is judged to be moderate-high). Some plant communities here may become stressed.
> 8) **St Road 520 to Three Forks Marsh (RK 378 - 442.8)**: This area is dominated by shallow marsh and includes Lake Poinsett. This is considered the uppermost region that could be affected by withdrawals; the likelihood of effects due to altered stage is judged to be moderate-high.
> 9) **Three Forks Marsh to Blue Cypress Lake (RK 442.8+)**: This segment will not be subject to water withdrawals.

four species, but it does not address the details of how these species might be used as indicators (i.e., what aspects of the populations will be measured, how will data be collected, what metrics might be developed?), or whether a field-based monitoring program is even planned. In addition, links between these populations and their predicted response to hydrologic change are not clearly presented. For example, a recent study by Darby et al. (2008) concluded that Florida apple snail (*Pomacea paludosa*) survival rates can be quite high in drought conditions, contrary to earlier reports of their low tolerance to drawdown. This calls to question the assumption that the snail kite can serve as an indicator of lowered water levels. Another concern is that many wetlands in the floodplain are not used as habitat by these avian species; therefore, these species may not provide much relevant information on wetland impacts. This is compounded if the bird populations occur in low densities, making it difficult to detect patterns in their population numbers or habitat use at the local scale.

The District reports high faunal diversity within the floodplain, including many species of fish, amphibians, reptiles, birds, mammals, and invertebrates. Many of these taxonomic groups are valuable indicators of hydrologic impacts to wetlands in monitoring programs (Micacchion,

2002) and might be considered as indicators in Phase II. For instance, reptiles and amphibians are valuable as monitoring targets because many exploit transition areas (e.g., wetland–upland) where the effects of dewatering may be seen. Amphibians are closely tied to wetland habitat. They depend on wetland pools for recruitment and serve as an important trophic link, preying upon insects and acting as a food source for predacious insects, larvae, fish, birds, and snakes. Amphibian-based indicators are effective assessment tools because of their sensitivity to even minor anthropogenic disturbance (Welsh and Ollivier, 1998; Adamus et al., 2001; Sparling et al., 2001).

Depending on its future use of indicators, the range of taxonomic groups used by the District in monitoring wetland impacts could be broadened. Ideas can be drawn from the considerable literature on the use of both plant and animal communities to assess wetland ecological condition (Sparling et al., 2001, Fennessy et al., 2002, Lopez and Fennessy, 2002). The district proposed investigating additional species in the Phase II work plans, but no details or methods were provided on how the different assemblages (e.g., reptiles, amphibians, etc.) might be used to indicate impacts. Rather than spending time on a literature review as proposed in Phase II, the District might consider field trials of existing biological indexes to determine their usefulness (for example, see Sparling et al., 2001). Many such indexes use multispecies assemblages as sources of interpretable signal that may be more robust than focusing on rare or other special interest species.

Data Availability and Analysis

Beyond the GIS data layers, very few data were presented in the wetland section of the Phase I report. As a result, it is difficult to understand how many of the conclusions of the wetlands workgroup were reached because of the many unstated assumptions embedded in the report. The report lacks details, partly because the Phase I hydrological modeling necessary for this analysis was still underway when the wetlands report was written. There are plans to address this in the Phase II study, although once again, few details are provided in the Phase II work plans on the methods that the district will use to accomplish this.

One concern is that the District appears to lack data against which future conditions can be compared. A strategic, field-based monitoring program to track key elements of the system that are expected to respond to hydrologic alterations could address this issue. Aside from the biological measures mentioned above, several chemical-physical measures could provide useful information. For example, to address concerns about the oxidation of organic soils due to dewatering, one approach is to monitor wetland subsidence or accretion using sediment elevation tables (Day et al., 1999, Cahoon et al., 2002). These measurements are both easy to make and, once equipment is installed, are not expensive.

An integrative method such as sediment elevation tables may provide information that is better able to reveal the impacts of hydrologic change than, for example, the soil analysis planned for Phase II. This proposed analysis involves the collection of soil samples at 150 locations throughout the basin to characterize the dominant soils in each river segment, particularly the organic soils vulnerable to drawdown. The approach is to perform a series of field and lab analyses on each sample to determine physical and chemical characteristics such as pH, conductivity, organic carbon and nutrient content, and concentrations of several metals. While these data may provide interesting insight into soil chemistry and will augment the work of the biogeochemistry workgroup, the committee has questions about (1) how these data will be incorporated into the GIS model, and (2) in conjunction with the biogeochemistry workgroup, how 150 soil cores, collected over the length of the St. Johns River, will be scaled up to provide information on the response of the system overall.

Additional Recommendations for Phase II

In addition to the recommendations provided above, several other issues need to be addressed in Phase II. One fundamental need is a complete methods section describing all assumptions, data collection techniques, analytical approaches, and metadata for the GIS layers.

Hydrologic modeling will be completed in Phase II, and the District plans to select a set of specific hydrological metrics believed to be most important for predicting vegetation community response. Indeed, the selection of these metrics should be one area where early integration of ideas and insights from multiple disciplines (in this case the hydrology and wetlands workgroups) is essential, as discussed in Chapter 2. Flooding depth, duration, and frequency, and water source, as well as water chemistry, are some key variables to consider. Additional analyses are needed to test the assumption that the effects of drawdown in the river propagate only a limited distance upslope in the wetlands and to identify how far upslope effects are felt (Doherty et al., 2000). This has important implications for predicting impacts to wetland functions such as providing spawning and nursery grounds for fish, amphibians and other species, and the maintaining of organic soils. It also can affect survival of vegetation in largely terrestrial areas near the edge of the floodplain. The committee is concerned that the landward edge of the floodplain may experience more dewatering than a 4-cm drop in river water level implies. How a drop in water level at Lake Jessup translates into the dewatering of adjacent floodplain and areas upstream should be addressed during Phase II.

The proposal to collect 150 additional soil samples should also be clarified to reflect that the biogeochemistry workgroup will be using these same soil cores in their analysis. This also speaks to the general need for better integration between the workgroups. For example, in the biogeochemistry sections of the Phase I report, soils are referred to as "floodplain" soils while the wetlands workgroup refers to "wetland" soils. The lack of consistent terminology has ramifications for the integration of the study as a whole and provides evidence that the groups are not communicating as much as they should.

The District has collected data on vegetation communities and elevation at a substantial number of floodplain transects throughout the basin. These data could be used as ground truth information to evaluate the GIS model output and verify the planned HGM classification of wetlands on the floodplain. Consideration could be also given to the use of these data as part of the Floristic Quality Assessment Index (FQAI), an index developed to assess the impact of human disturbance on vegetation communities. This index, akin to a plant-based Index of Biotic Integrity, assigns a repeatable and quantitative value to vegetation community composition and could be used to detect shifts in vegetation community composition that might occur with water withdrawals. Use of the index requires that the local flora be identified (including invasive species) and that the coefficients needed for each species be available; fortunately, these coefficients are available for Florida (Cohen et al., 2004). The FQAI has been shown to be an effective plant-based bioassessment tool in many regions (Lopez and Fennessy, 2002) including

Florida (Cohen et al., 2004), and it could prove useful if the District intends to monitor wetlands to track wetland response to management activities.

Summary

In order to move beyond the descriptive Phase I work, the wetlands workgroup will need to enhance the methods used to identify and delineate wetlands that are influenced by the river's hydrology. Fine-scale elevation data in the form of a LiDAR-based DEM are needed to produce accurate maps in flat, wetland-rich areas, particularly in efforts to characterize hydrology and map wetland diversity. The District plans during Phase II to produce DEMs for the portions of the watershed where data are available. It will also be important for the workgroup to better define the criteria used to determine the likelihood of impacts of water withdrawals to each wetland segment. Finally, the District should consider broadening the range of taxonomic groups used in monitoring wetland impacts. Amphibians, reptiles, invertebrates, and plants have proven to be valuable indicators of hydrologic impacts to wetlands in monitoring programs and might be considered as indicators in Phase II.

References

Adamus, P., T. J. Danielson, and A. Gonyaw. 2001. Indicators for monitoring biological integrity of inland, freshwater wetlands. Grant # 843-R-01-xx. Washington D.C.: EPA Office of Wetlands, Oceans, and Watersheds.

Allen, D. M., and D. L. Barker. 1990. Interannual variations in larval fish recruitment to estuarine epibenthic habitats. Marine Ecology Progress Series 63:113–125.

Alonso-Saez, L., E. Vazquez-Dominguez, C. Cardelus, J. Pinhassi, M. M. Sala, I. Lekunberri, V. Balague, M. Vila-Costa, F. Unrein, R. Massana, R. Simo, and J. M. Gasol. 2008. Factors controlling the year-round variability in carbon flux through bacteria in a coastal marine system. Ecosystems 11:397–409.

Batiuk, R. A., R. J. Orth, K. A. Moore, W. C. Dennison, J. C. Stevenson. 1992. Chesapeake Bay Submerged Aquatic Vegetation Habitat Requirements and Restoration Goals: A Technical Synthesis. Annapolis, MD: EPA Chesapeake Bay Program.

Beck, M. W., K. L. Heck, K. W. Able, D. L. Childers, D. B. Eggleston, B. M. Gillanders, B. Halpern, C. G. Hays, K. Hoshino, and T. J. Minello. 2001. The identification, conservation, and management of estuarine and marine nurseries for fish and invertebrates. BioScience 51:633–641.

Biber, D. C., C. L. Gallegos, and W. J. Kenworthy. 2008. Calibration of a bio-optical model in the North River, North Carolina (Albemarle-Pamlico Sound): A tool to evaluate water quality impacts on seagrasses. Estuaries and Coasts 31:177–191.

Bilkovic, D. M., C. H. Hershner, M. R. Berman, K. J. Havens, and D. M. Stanhope. 2005. Evaluating nearshore communities as indicators of ecosystem health. Pp. 365–379 *In* Estuarine Indicators. S. Bortone (ed.). Boca Raton, FL: CRC Press.

Blocksom, K. A. 2003. A performance comparison of metric scoring methods for a multimetric index for Mid-Atlantic Highlands streams. Environmental Management 31:670–682.

Bonada, N., N. Prat, V. H. Resh, and B. Statzner. 2006. Developments in aquatic insect biomonitoring: A comparative analysis of recent approaches. Annual Review of Entomology 51:495–523.

Booth, D. B., J. R. Karr, S. Schauman, K. P. Konrad, S. A. Morley, M. G. Larson, and S. J. Burges. 2004. Reviving urban streams: land use, hydrology, biology, and human behavior. Journal American Water Resources Association 40:1351–1364.

Borja, A., S. B. Bricker, D. M. Dauer, N. T. Demetriades, J. G. Ferreira, A. T. Forbes, P. Hutchings, X. Jia, R. Kenchington, J. C. Marques, and C. Zhu. 2008. Overview of integrative tools and methods in assessing ecological integrity in estuarine and coastal systems worldwide. Marine Pollution Bulletin 56:1519–1537.

Bradley, P., W. S. Fisher, H. Bell, W. S. Davis, V. Chan, C. LoBue, and W. Wiltse. 2008. Development and implementation of coral reef biocriteria in U. S. jurisdictions. Environmental Monitoring and Assessment 150:43–51.

Breder, C. M., Jr., and D. E. Rosen. 1966. Modes of Reproduction in Fishes. Garden City, NY: The Natural History Press.

Brinson, M. M. 1993. A hydrogeomorphic classification for wetlands. Technical Report WRP-DE-4. NTIS No. AD A270 053. Vicksburg, MS: U.S. Army Engineer Waterways Experiment Station.

Burton & Associates. 2004. Affordability analysis of alternative water supply, St. Johns River Water Supply Project. Special Publication SJ2004-SP27. Palatka, FL: St. Johns River Water Management District.

Butler, D. E., H. A. Radin, S. Sunderland, A. Montoya, and B. E. McGurk. 2009. East Central Florida Transient Model Documentation. West Palm Beach, FL: South Florida Water Management District and St. Johns River Water Management District.

Cahoon, D. R., J. C. Lynch, B. C. Perez, B. Segura, R. D. Holland, C. Stelly, G. Stephenson and P. Hensel. 2002. High-Precision Measurements of Wetland Sediment Elevation: II. The Rod Surface Elevation Table. Journal of Sedimentary Research 72:734–739.

Carpenter, S. 1996. Microcosm experiments have limited relevance for community and ecosystem ecology. Ecology 77:677–680.

Chen, C.-C., J. E. Petersen, and W. M. Kemp. 2000. Nutrient uptake in experimental estuarine ecosystems: scaling and partitioning rates. Marine Ecological Progress Series 200:103–116.

CH2M Hill. 1996a. Water Supply Needs and Sources Assessment, Alternative Water Supply Strategies Investigation, Surface Water Data Acquisition and Evaluation Methodology. Special Publication SJ96-SP1, Gainesville, FL.

CH2M Hill. 1996b. Water Supply Needs and Sources Assessment, Alternative Water Supply Strategies Investigation, Surface Water Withdrawal Sites. Special Publication SJ96-SP4, Gainesville FL.

CH2M Hill. 1997. Water Supply Needs and Sources Assessment, Alternative Water Supply Strategies Investigation, Surface Water Availability and Yield Analysis. Special Publication SJ97-SP7. Palatka FL: St. Johns River Water Management District.

CH2M Hill. 2004. Surface water treatability and demineralization study, St. Johns River Water Supply Project. Special Publication SJ2004-SP20. Palatka, FL: St. Johns River Water Management District.

Cohen, M. J., S. Carstenn, and C. R. Lane. 2004. Floristic Quality Indices for Biotic Assessment of Depressional Marsh Condition in Florida. Ecological Applications 14(3):784–794.

Conley, D. 2000. Biogeochemical nutrient cycles and nutrient management strategies. Hydrobiologia 410:87–96.

Darby, P. C., R. E. Bennetts, and H. F. Percival. 2008. Dry down impacts on apple snail (Pomacea paludosa) demography: Implications for wetland water management. Wetlands 28:204–214.

Davies, S. P., and S. K. Jackson. 2006. The biological condition gradient: a descriptive model for interpreting change in aquatic ecosystems. Ecological Applications 16:1251–1266.

Davis, J. L. D., L. A. Levin, and S. M. Walter. 2002. Artificial armored shorelines: sites for open-coast species in a southern California bay. Marine Biology 140:1249–1262.

Day, J. W., J. Rybczyk, F. Scarton, A. Rismondo, D. Are, and G. Cecconi. 1999. Soil Accretionary Dynamics, Sea-Level Rise and the Survival of Wetlands in Venice Lagoon: A Field and Modelling Approach. Estuarine, Coastal and Shelf Science 49: 607–628.

Deegan, L. A., J. T. Finn, and J. Buonaccorsi. 1997. Development and validation of an estuarine biotic integrity index. Estuaries 20:601–617.

DeLong, A. K. and J. S. Collie. 2004. Defining essential fish habitat: a model-based approach. Narragansett, RI: Rhode Island Sea Grant.

Doberfuhl, D. R. 2007. Light limiting thresholds for submerged aquatic vegetation in a blackwater river. Aquatic Botany 86:346–352.

Doherty, S., M. Cohen, C. Lane, L. Line, and J. Surdick. 2000. Biological Criteria for Inland Freshwater Wetlands in Florida: A Review of Technical & Scientific Literature (1990-1999). A Report to the United States Environmental Protection Agency. Center for Wetlands, University of Florida, Gainesville, FL 32611.

EPA (U.S. Environmental Protection Agency). 2003. Draft Report on the Environment 2003. EPA-260-R-02-006. Washington, DC: EPA Office of Environmental Information and the Office of Research and Development.

Fennessy, M. S., M. Gernes, J. Mack, and D. Heller-Wardrop. 2002. Using vegetation to assess environmental conditions in wetlands. EPA 843-B-00-0002j. Washington, DC: EPA Office of Water.

FFWCC (Florida Fish and Wildlife Conservation Commission). 2001. Fisheries-Independent monitoring program 2001 annual data summary. Jacksonville, FL: Fish and Wildlife Research Institute.

FDEP (Florida Department of Environmental Protection). 2009. 2007 Reuse Inventory. Available online at http://www.dep.state.fl.us/water/reuse/inventory.htm.

Fisher, W. S., L. S. Fore, A. Hutchins, R. L. Quarles, J. G. Campbell, C. LoBue, and W. S. Davis. 2008. Evaluation of stony coral indicators for coral reef management. Marine Pollution Bulletin 56:1737–1745.

Fore, L. S. 2003. Developing Biological Indicators: Lessons Learned from Mid-Atlantic Streams. EPA-903-R-003. Ft. Meade, MD: EPA Office of Environmental Information and Mid-Atlantic Integrated Assessment Program, Region 3. (www.epa.gov/bioindicators).

Fore, L., R. Frydenborg, D. Miller, T. Frick, D. Whiting, J. Espy, and L. Wolfe. 2007. Development and Testing of Biomonitoring Tools for Macroinvertebrates in Florida Streams (Stream Condition Index and Biorecon). Report prepared for Florida Dept. of Environmental Protection, Tallahassee, FL. Available online at ftp://ftp.dep.state.fl.us/pub/labs/assessment/sopdoc/sci_final.pdf.

Fredrette, T. J., R. J. Diaz, J. Monfrans and R. Orth. 1990. Secondary production within a seagrass bed (*Zostera marina* and *Ruppia maritima*) in lower Chesapeake Bay. Estuaries 13:431–440.

Freeman, M. C., and P. A. Marcinek. 2006. Fish assemblage responses to water withdrawals and water supply reservoirs in piedmont streams. Environmental Management 38:435–450.

Fowler-Walker, M. J., S. D. Connell, and B. M. Gillanders. 2005. Variation at local scales need not impede tests for broader scale patterns. Marine Biology 147:823–831.

Gray, C. A. 1996. Small-scale temporal variability in assemblages of larval fishes: implications for sampling. Journal of Plankton Research 18:1643–1657.

Hagan, E. R., K. J. Holmes, J. E. Kiang, and R. C. Steiner. 2005. Benefits of Iterative Water Supply Forecasting in the Washington, D.C., Metropolitan Area. Journal of the American Water Resources Association (JAWRA) 41(6):1417–1430.

Hall, G. 2005. Ocklawaha River water allocation study, Technical Publication SJ2005-1. Palatka, FL: St. Johns River Water Management District.

Harbaugh, A. W., E. R. Banta, M. C. Hill, and M. G. McDonald. 2000. MODFLOW-2000, the U.S. Geological Survey Modular Ground-water Model – User Guide to Modularization Concepts and the Ground-Water Flow Process: U.S. Geological Survey Open-File Report 00-92. Washington, DC: U.S. Geological Survey.

Houde, E. D. and J. D. A. Lovdal. 1985. Patterns of variability in ichthyoplankton occurrence and abundance in Biscayne Bay, Florida. Estuarine, Coastal and Shelf Science 20:79–103.

Houston, L., M. T. Barbourb, D. Lenatc, and D. Penrose. 2002. A multi-agency comparison of aquatic macroinvertebrate-based stream bioassessment methodologies. Ecological Indicators 1(4): 279–292.

Hughes, J. E., L. A. Deegan, M. J. Weaver, and J. E Costa. 2002. Regional application of and index of estuarine biotic integrity based on fish communities. Estuaries 25:250–263.

Hyde, P., R. Dubayah. W. Walker, J. B. Blair, M. Hofton, and C. Hunsaker. 2006. Mapping forest structure for wildlife habitat analysis using multi-sensor (LiDAR, SAR/InSAR, ETM+, Quickbird) synergy. Remote Sensing of Environment 102(1–2):63–73.

Joint, I., P. Henriksen, G. A. Fonnes, D. Bourne, T. F. Thingstad, and B. Riemann. 2002. Competition for inorganic nutrients between phytoplankton and bacterioplankton in nutrient manipulated mesocosms. Aquatic Microbial Ecology 29:149–159.

Jordan, S. J., and P. A. Vass. 2000. An index of ecosystem integrity for Northern Chesapeake Bay. Environmental Science and Policy 3:S59–S88.

Karr, J. R. 2006. Seven foundations of biological monitoring and assessment. Biologia Ambientale 20(2):7–18.

Keast, A. 1985. The piscivore feeding guild of fishes in small freshwater ecosystems. Environmental Biology of Fishes 12(2):119–129.

Keast, A., and J. Eadie. 1984. Growth in the first summer of like: a comparison of nine co-occurring fish species. Canadian Journal of Zoology 62:1242–1250.

Keast, A., J. Harker, and D. Turnbull. 1978. Nearshore fish habitat utilization and species associations in Lake Opinicon (Ontario, Canada). Environmental Biology of Fishes 3(2):173–184.

Keeler, A. G., and D. McLemore. 1996. The value of incorporating bioindicators in economic approaches to water pollution. Ecological Economics 19: 237–245.

Kemp, W. M., W. R. Boynton, J. E. Adolf, D. F. Boesch, W. C. Boicourt, G. Brush, J. C. Cornwell, T. R. Fisher, P. M. Glibert, J. D. Hagey, L .R. Harding, E. D. Houde, D. G. Kimmel, W. D. Miller, R. I. E. Newell, M. R. Roman, E. M. Smith and J. C. Stevenson. 2005. Eutrophication of Chesapeake Bay: Historical trends and ecological interactions. Marine Ecology Progress Series 303:1–29.

Kimmerer, W. J., E. S. Gross, and M. L. MacWilliams. 2009. Is the response of estuarine nekton to freshwater flow in the San Francisco estuary explained by variation in habitat volumes? Estuaries and Coasts 32:375–389.

Kirk, J. T. O. 1994. Light and Photosynthesis in Aquatic Ecosystems, 2nd ed. Cambridge, MA: Cambridge University Press.

Klemm, D. J., K. A. Blocksom, F. A. Fulk, A. T. Herlihy, R. M. Hughes, P. R. Kaufmann, D. V. Peck, J. L. Stoddard, W. T. Thoeny, M. B. Griffith, and W. S. Davis. 2003. Development and evaluation of a macroinvertebrate biotic integrity index (MBII) for regionally assessing Mid-Atlantic Highland streams. Environmental Management

31(5):656–669.

Klemm, D. J., K. A. Blocksom, W. T. Thoeny, F. A. Fulk, A. T. Herlihy, P. R. Kaufmann, and S. M. Cormier. 2002. Using macroinvertebrates as indicators of ecological conditions for streams in the Mid-Atlantic Highlands region. Environmental Monitoring and Assessment 78:169–212.

Kovalenko, K., E. D. Dibble, and R. Fugi. 2009. Fish feeding in changing habitats: effects of invasive macrophyte control and habitat complexity. Ecology of Freshwater Fish 18(2):305-313.

Lang, M. W. and G. W. McCarty. 2008. Innovative tools for mapping forested wetlands in the Choptank River Watershed. Maryland Natural Resources Conservation Service. Available at: http://www.md.nrcs.usda.gov/technical/ceap.html.

Larsen, D. P., T. M. Kincaid, S. E. Jacobs, and N. S. Urquhart. 2001. Design for evaluating local and regional-scale trends. BioScience 51:1069–1078.

Lopez, R., and M. S. Fennessy. 2002. Testing the Floristic Quality Assessment Index as an indicator of wetland condition along gradients of human influence. Ecological Applications 12:487–497.

Lowery, T. 1998. Modelling estuarine eutrophication in the context of hypoxia, nitrogen loadings, stratification and nutrient ratios. Journal of Environmental Management 52:289–305.

MacArthur, R. H. 1965. Patterns of species diversity. The Biological Bulletin 40:510–533.

Mace, J. W. 2006. Minimum Levels Determination: St. Johns River at State Road 44 near DeLand, Volusia County, Technical Publication SJ2006-5. Palatka, FL: St. Johns River Water Management District.

Maxa, M., and P. Bolstad. 2009. Mapping northern wetlands with high resolution satellite images and LiDAR. Wetlands 29:248–260.

McGurk, B., and P. Presley. 2002. Simulation of the Effects of Groundwater Withdrawals on the Floridan Aquifer System in East-Central Florida: Model Expansion and Revision. Technical Publication SJ2003-3. Palatka, FL: St. Johns River Water Management District.

Micacchion, M. 2002. Amphibian Index of Biotic Integrity (AmphIBI) for Wetlands. Final Report to U.S. EPA Grant No. CD985875-01 Volume 3. Columbus, OH: Ohio Environmental Protection Agency, Division of Surface Water, Wetland Ecology Group.

Moody, H. R. 1961. Exploited fish populations of the St. Johns River, Florida. The Quarterly Journal of the Florida Academy of Sciences 24(1):1–18.

Morse, W. W. 1989. Catchability, growth, and mortality of larval fishes. Fishery Bulletin 87:417–446.

Motz, L. H., and A. Dogan. 2004. North-Central Florida Active Water-Table Regional Groundwater Flow Model (Final Report). Contract Number 99W384/ UF450472612. Palatka, FL: St. Johns River Water Management District.

Neubauer, C. P., G. B. Hall, E. F. Lowe, C. Robinson, R. Hupalo, and L. Keenan. 2008. Minimum Flows and Levels Method of the St. Johns River Water Management District, Florida, USA. Environmental Management 42:1101-1114.

Niemi, G. J., and M. E. McDonald. 2004. Application of ecological indicators. Annual Review of Ecology and Systematics 35:89–111.

Niemi, G. J., D. H. Wardrup, R. P. Brooks, S. Anderson, V. Brady, and H. Paerl. 2004. Rationale for a new generation of indicators for coastal waters. Environmental Health

Perspectives 112:979–986.

Norton, S. B., S. M. Cormier, M. Smith, and R. Christian Jones. 2000. Can biological assessments discriminate among types of stress? A case study from the Eastern Corn Belt Plains ecoregion. Environmental Toxicology and Chemistry 19:1113–1119.

Paerl, H. W. 2009. Controlling eutrophication along the freshwater–marine continuum: Dual nutrient (N and P) reductions are essential. Estuaries and Coasts DOI 10.1007/s12237-009-9158-8.

Peebles, E. B., S. E. Burghardt, and D. J. Hollander. 2007. Causes of interestuarine variability in bay anchovy (*anchoa mitchilli*) salinity at capture. Estuaries and Coasts 30:1060–1074.

Peterson, M. S. 1988. Comparative physiological ecology of centrarchids in hyposaline environments. Canadian Journal of Fisheries and Aquatic Sciences 45(5):827–833.

Peterson, M. S. 2003. A conceptual view of environmental-habitat-production linkages in tidal river estuaries. Reviews in Fisheries Science 11:291–313.

Peterson, M. S., and S. T. Ross. 1991. Dynamics of littoral fishes and decapods along a coastal river-estuarine gradient. Estuarine, Coastal & Shelf Science 33(5):467–483.

Peterson, M. S., and M. R. Meador. 1994. Effects of salinity on freshwater fishes in coastal plain drainages in the southeastern United States. Reviews in Fisheries Science 2(2):95–121.

Peterson, M. S., and S. J. VanderKooy. 1995. Phenology and spatial and temporal distribution of larval fishes in a partially channelized warmwater stream. Ecology of Freshwater Fish 4:93–105.

Peterson, M. S., M. R. Weber, M. L. Partyka, and S. T. Ross. 2007. Integrating *in situ* quantitative geographic information tools and size-specific laboratory-based growth zones in a dynamic river-mouth estuary. Aquatic Conservation: Marine and Freshwater Ecology 17(6):602–618.

Pont, D., B. Hugueny, and C. Rogers. 2007. Development of a fish-based index for assessment of river health in Europe: the European fish index. Fisheries Management and Ecology 14:427–439.

Pont, D., R. M. Hughes, T. R. Whittier, and S. Schmutz. 2009. A predictive index of biotic integrity model for aquatic-vertebrate assemblages of western U.S. streams. Transactions of the American Fisheries Society 138:292–305.

Porter, E., L. P. Sanford, G. Gust, and F. S. Porter. 2004. Combined water-column mixing and benthic boundary-layer flow in mesocosms: key for realistic benthic-pelagic coupling studies. Marine Ecological Progress Series 271:43–60.

Rahmstorf, S. 2007. A semi-empirical approach to projecting future sea-level rise. Science 315:368–370.

Rheinhardt, R. D., M. M. Brinson, P. M. Farley. 1997. Applying wetland reference data to functional assessment, mitigation, and restoration. Wetlands 17:195–215.

Robison, C. P. 2004. Middle St. Johns River Minimum Flows and Levels Hydrologic Methods Report, Technical Publication SJ2004-2. Palatka, FL: St. Johns River Water Management District.

Sagan, J. J. 2007. SAV Monitoring Project: Interim Reports I-V. Interim reports associated with quarterly sampling and ground truthing surveys for the St. Johns River Water Management District, Palatka FL.

Schultz, G. E. H. D. 2000. Changes in bacterioplankton metabolic capabilities along a salinity gradient in the York River estuary, Virginia, USA. Aquatic Microbial Ecology 22:163–174.

Shaw, R. F., D. L. Drullinger, K. A. Edds, and D. L. Leffler. 1988. Fine-scale spatial distribution of red drum, *Sciaenops ocellatus*, larvae. Contributions in Marine Science 30:109–116.

Simenstad, C. A., S. B. Brandt, A. Chambers, R. Dame, L. A. Deegan, R. Hodson, and E. D. Houde. 2000. Habitat–biotic interactions. *In* Estuarine Science: A Synthetic Approach to Research and Practice. Hobbie J. E. (ed.). Washington, D.C.: Island Press.

Solomon, J. J., and R. B. Brodie. 2007. Fisheries-independent monitoring program-Independent data from expansion sampling on the St. Johns River July 2006–June 2007. St. Petersburg FL: Florida Fish and Wildlife Conservation Commission, Marine Research Institute.

Sparling, D. W., K. O. Richter, A. Calhoun, and M. Micacchion. 2001. Methods for evaluating wetland condition: using amphibians in bioassessments of wetlands. EPA 822-R-01-0071. Washington D.C.: EPA Office of Water.

SJRWMD (St. Johns River Water Management District). 2000. District Water Supply Plan, 2000. Special Publication SJ2000-SP1. B. A. Vergara (ed.). Palatka, FL: St. Johns River Water Management District.

SJRWMD. 2006. District Water Supply Plan, 2005. Technical Publication SJ2006-2. Palatka, FL: St. Johns River Water Management District.

SJRWMD. 2008. Alternative Water Supply Cumulative Impact Assessment, Interim Report-Draft. Lowe, E. F., L. E. Battoe, and T. Bartol (eds.). Palatka, FL: St. Johns River Water Management District.

SJRWMD. 2009a. Water Supply Assessment 2008 (Draft). Technical Publication SJ2009-_(in preparation). Palatka, FL: St Johns River Water Management District.

SJRWMD. 2009b. District Water Supply Plan, 2005, Fourth Addendum. Technical Publication SJ2006-2D. Palatka, FL: St Johns River Water Management District.

SJRWMD. 2009c. Methods for the Second Phase of the Water Supply Study For Review by the NRC Panel.

Sokal, R. R., and F. J. Rohlf. 1995. The principles and practice of statistics in biological research. 3rd Edition. New York: Freeman and Company.

Staver, L. W., K. W. Staver, and J. C. Stevenson. 1996. Nutrient inputs to the Choptank River Estuary: Implications for watershed management. Estuaries 19:342–358.

Stevenson, J. C., and N. Confer. 1978. Summary of Available Information on Chesapeake Bay Submerged Vegetation. U.S. Dept. Interior, Fish and Wildlife Service, Biological Services Program (FWS/OBS-78/66) NTIS.

Stevenson, J. C., L. W. Staver, and K. Staver. 1993. Water quality associated with survival of submersed aquatic vegetation along an estuarine gradient. Estuaries 16:346–361.

Stoddard, J. L., D. P. Larsen, C. P. Hawkins, R. K. Johnson, and R. H. Norris. 2006. Setting expectations for the ecological condition of streams: the concept of reference condition. Ecological Applications 16:1267–1276.

Theel, H. J., E. D. Dibble, and J. D. Madsen. 2007. Differential influence of a monotypic and diverse native aquatic plant bed on a macroinvertebrate assemblage; an experimental implication of exotic plant induced habitat. Hydrobiologia 600(1):77–87.

Titus, J. G., and V. Narayanan. 1995. The Probability of Sea Level Rise. EPA 230-R95-008. Washington, D.C.: EPA.

References

Tranvik, L. 1990. Bacterioplankton growth on fractions of dissolved organic carbon of different molecular weights from humic and clear waters. Applied Environmental Microbiology 56:1672–1677.

Vannote, R. L., G. W. Minshall, K. W. Cummins, J. R. Sedell, and C. E. Cushing. 1980. The River Continuum Concept. Canadian Journal of Fisheries and Aquatic Sciences 37(1):130–137.

Wagner, C. M. 1999. Expression of the estuarine species minimum in littoral fish assemblages of the lower Chesapeake Bay tributaries. Estuaries 22(2A): 304–312.

Wagner, C. M., and H. M. Austin. 1999. Correspondence between environmental gradients and summer littoral fish assemblages in low salinity reaches of the Chesapeake Bay, USA. Marine Ecology Progress Series 177: 197–212.

Walburg, C. H. 1960. Abundance and life history of shad St. Johns River, Florida. Fishery Bulletin 60:486–501.

Wardrup, D. H., M. E. Kentula, D. L. Stevens Jr., S. F. Jensen, and R. P. Brooks. 2007. Assessment of wetland condition: an example from the Upper Juniata Watershed in Pennsylvania, USA. Wetlands 27:416–431.

Wardrup, D. H., C. Hershner, K. Havens, K. Thornton, and D. M. Bilkovic. 2007. Developing and communicating a taxonomy of ecological indicators: A case study from the Mid-Atlantic. EcoHealth 4:179–186.

Welsch, H. H., and L. M. Ollivier. 1998. Stream amphibians as indicators of ecosystem stress: a case study from California redwoods. Ecological Applications 8:1118–1132.

Whigham, D. L. 1999. Ecological issues related to wetland preservation, restoration, creation and assessment. The Science of the Total Environment 240:31–40.

Wigand, C., J. C. Stevenson, and J. C. Cornwell. 1997. Varying sediment biogeochemistry of submersed macrophytes of the tidal freshwater Upper Chesapeake Bay. Aquatic Botany 56:233–244.

Yoder, C. O., and J. E. DeShon. 2003. Using biological response signatures within a framework of multiple indicators to assess and diagnose causes and sources of impairments to aquatic assemblages in selected Ohio rivers and streams. Pp. 23–81 *In:* Biological response signatures: indicator patterns using aquatic communities. T. P. Simon (ed.). Boca Raton, FL: CRC Press.

Yoder, C. O., and M. T. Barbour. 2009. Critical technical elements of state bioassessment programs: a process to evaluate program rigor and comparability. Environmental Monitoring and Assessment 150:31–42.

Appendix A

Acronyms

CDOM	colored dissolved organic matter
CFCA	Central Florida Coordination Area
CH3D	Curvilinear Hydrodynamics 3-Dimensional Model
CIA	Cumulative Impact Assessment
DEM	digital elevation model
DOC	dissolved organic carbon
EAA	Everglades Agricultural Area
ECF	East Central Florida Model
ECFT	East Central Florida Transient Model
EFDC	Environmental Fluid Dynamics Model
FFWCC	Florida Fish and Wildlife Conservation Commission
FIM	Fisheries Independent Monitoring
FQAI	Floristic Quality Assessment Index
GIS	geographic information systems
HGM	Hydrogeomorphic Approach
HSPF	Hydrologic Simulation Program-Fortran
LIDAR	Light Detection and Ranging
MFL	minimum flow and level
MGD	million gallons per day
MODFLOW	modular finite-difference flow model
NCF	North Central Florida Model
NRC	National Research Council
PWRCA	Priority Water Resource Caution Areas
RK	river kilometer
SAS	surficial aquifer system
SAV	submersed aquatic vegetation

SJRWMD	St. Johns River Water Management District
TMDL	Total Maximum Daily Load
UFA	upper Floridan aquifer
USGS	U.S. Geological Survey
WSA	Water Supply Assessment
WSIS	Water Supply Impact Study

Appendix B

Biographical Sketches for Committee to Review the St. Johns River Water Supply Impact Study

Patrick L. Brezonik, *Chair,* is professor and Fesler-Lampert Chair of Urban and Regional Affairs in the Department of Civil Engineering at the University of Minnesota. His research interests are focused on the impacts of human activity on water quality and the biogeochemical cycles of important elements (nitrogen, phosphorus, trace metals) in large natural aquatic systems. Field studies, including experimental manipulations in large systems, and modeling approaches are emphasized. Dr. Brezonik is a former program director for environmental engineering at the National Science Foundation and has served on numerous NRC committees, including the Committee to Review the Corps of Engineers Restructured Upper Mississippi River-Illinois Water Draft Feasibility Study, the Committee on Restoration of the Greater Everglades Ecosystem, and is a past member of the Water Science and Technology Board. He received his B.S. in chemistry from Marquette University, and his M.S. and Ph.D. in water chemistry from the University of Wisconsin-Madison.

M. Siobhan Fennessy is a professor of biology at Kenyon College. Her areas of expertise are in aquatic ecology, wetland plant community dynamics, and landscape ecology. Dr. Fennessy's primary areas of research are freshwater ecosystems, their plant communities and restoration, how ecosystems respond to human impacts, and the role of temperate wetlands in the global carbon cycle. She previously served on the faculty of the Geography Department of University College London and held a joint appointment at the Station Biologique du la Tour du Valat (located in southern France) where she conducted research on human impacts to Mediterranean wetlands. She recently co-authored a book on the ecology of wetland plants. Dr. Fennessy is a 2001 recipient of the Robert J. Tomsich Science Award for excellence in scientific research. She received her B.S. in botany and Ph.D. in environmental science from the Ohio State University.

Ben R. Hodges is an associate professor of civil engineering at the University of Texas at Austin. His primary areas of interest are in the fields of environmental fluid mechanics and surface water hydraulics; coupled field and model investigations of hydrodynamics in lakes, rivers, and estuaries; relationships between river hydraulics and instream flow for aquatic habitat; and linkages between water quality and hydrodynamics in natural systems. His recent research has focused on hydrodynamic and transport modeling of the stratification in Corpus Christi Bay, which impacts episodic hypoxia development. Also, Dr. Hodges is familiar with Florida river systems through his outside peer review of the Lower Peace River and Shell Creek modeling for minimum flow requirements for the Southwest Florida Water Management District. He received his B.S. in marine engineering and nautical science from the U.S. Merchant Marine Academy, his M.S. in mechanical engineering from George Washington University, and his Ph.D. in civil engineering from Stanford University.

Appendix B

James R. Karr is Professor Emeritus of Ecology and Environmental Policy at the University of Washington, Seattle. His primary areas of interest span from tropical forest ecology to aquatic ecology and watershed management, with a specific focus on fostering use of ecological knowledge to resolve complex natural resource and environmental disputes. Dr. Karr's recent research has focused on the ecology of fish and invertebrates in streams, on plants and invertebrates in shrub steppe, and the demography of tropical forest birds. He has served on numerous review teams for both the South Florida Water Management District and the Florida Department of Environmental Protection. He served on the NRC Committee to Assess the Scientific Basis of the Total Maximum Daily Load Approach to Water Pollution Reduction. Dr. Karr received his B.Sc. in fish and wildlife biology from Iowa State University and his M.Sc. and Ph.D. degrees in zoology from the University of Illinois.

Mark S. Peterson is a professor in the Department of Coastal Sciences at the University of Southern Mississippi. His primary areas of expertise are in fisheries ecology, population biology, sustainable coastal development, habitat loss, invasive species, and Geographic Information Systems (GIS). Dr. Peterson's current research interests are broadly focused on habitat use in nekton, with particular emphasis on factors affecting recruitment success and distribution in estuarine-dependent fishes and the tradeoffs made by nekton when living in different habitats. He has also recently begun addressing habitat loss and environmental variability and impacts on fish recruitment and distribution using GIS. He served on the NRC Committee to Review the Florida Keys Carrying Capacity Study. He received his B.S. in marine science from Coastal Carolina University, his M.S. in bioenvironmental oceanography, and his Ph.D. in biological sciences from the University of Southern Mississippi.

James L. Pinckney is a professor of marine and biological sciences at the University of South Carolina. His research interests are focused on marine and microbial ecology, microalgal ecophysiology, phytoplankton–nutrient interactions, harmful algal blooms, and ecosystem eutrophication in estuarine and coastal habitats of Texas. His specific interests are centered around the ecophysiological factors and processes that influence carbon partitioning, allocation, and interspecific competition in multispecies assemblages. Dr. Pinckney's current research is being conducted in local estuarine, riverine, and intertidal habitats, as well as in pelagic systems in the Gulf of Mexico and hypersaline lagoons in the Bahamas. He received his B.S. in biology and his M.S. in marine biology from the College of Charleston, and his Ph.D. in ecology from the University of South Carolina.

Jorge I. Restrepo is a professor of geohydrology and director of the Hydrological Modeling Center of the Department of Geography and Geology at Florida Atlantic University. His current research interests include evapotranspiration in southern Florida; modeling recharge, evapotranspiration, and runoff; development of a wetland simulation model; modeling of seepage in the Everglades Nutrient Removal Site Test Cells; development of a generalized computer model to represent physical and operational behavior of a stream–aquifer system for evaluating conjunctive management of surface water and groundwater; and development of an optimization model to support the planning of a regional aquifer storage and recovery facility along a canal system. Dr. Restrepo served on the NRC Committee on Sustainable Underground

Storage of Recoverable Water. He received his B.A. from the Universidad Nacional, Facultad de Minas, and his M.A. and Ph.D. from Colorado State University.

Roland C. Steiner is the regional water and wastewater manager for the Washington Suburban Sanitary Commission (WSSC). He is responsible for developing and maintaining the major functional and financial relationships between WSSC and adjacent utilities, including management and funding of cost-shared water supply reservoirs and advisory services, water curtailment agreements, and reconciliation of WSSC's capital funding at several wastewater treatment plants. Previously he was associate director for Water Resources and Director of Cooperative Water Supply Operations on the Potomac for the Interstate Commission on the Potomac River Basin. There he was responsible for directing the water resources program of the Commission including covering coordinated drought supply management of river and reservoir resources for the Washington, DC region. He is a licensed professional engineer in Maryland. He served on the NRC Committee on Water Resources Activities at the U.S. Geological Survey. He has a B.Sc. in civil engineering from the University of Pennsylvania, and M.Sc. in civil engineering from Stanford University, and a Ph.D. in environmental engineering from the Johns Hopkins University.

J. Court Stevenson is a professor at the Horn Point Laboratory of the University of Maryland Center for Environmental Science. His primary areas of interest are coastal zone resources and water quality management issues; ecology of marsh and sea grass systems; effects of sea-level rise on wetlands and coastal shorelines; and the environmental history of Chesapeake Bay and its watershed. Dr. Stevenson received his B.S. in biology from Brooklyn College of the City University of New York, and his Ph.D. in botany from the University of North Carolina at Chapel Hill.

Laura J. Ehlers is a senior staff officer for the Water Science and Technology Board of the National Research Council. Since joining the NRC in 1997, she has served as the study director for 13 committees, including the Committee to Review the New York City Watershed Management Strategy, the Committee on Bioavailability of Contaminants in Soils and Sediment, the Committee on Assessment of Water Resources Research, and the Committee on Reducing Stormwater Discharge Contributions to Water Pollution. Ehlers has periodically consulted for EPA's Office of Research Development regarding their water quality research programs. She received her B.S. from the California Institute of Technology, majoring in biology and engineering and applied science. She earned both an M.S.E. and a Ph.D. in environmental engineering from the Johns Hopkins University.